# Strategic Business Decisions

R. Srinivasan

# Strategic Business Decisions

## A Quantitative Approach

Springer

R. Srinivasan
Department of Management Studies
Indian Institute of Science
Bangalore, Karnataka
India

ISBN 978-81-322-1900-2 ISBN 978-81-322-1901-9 (eBook)
DOI 10.1007/978-81-322-1901-9
Springer New Delhi Heidelberg New York Dordrecht London

Library of Congress Control Number: 2014936866



Printed on acid-free paper

Springer is part of Springer Science+Business Media (www.springer.com)

*At the Lotus Feet*
*of the Lord of seven hills*
*Lord Shri Venkateshwara*

# Preface

*Strategic Business Decisions* is an attempt to outline the essentials of Operations Research and Engineering Management and their applicability for effective decision making in a manner that is easily understood by graduate engineering students and postgraduate students of management. The hands-on approach used throughout the book and the practical nature of the illustrations and case studies discussed make the book a handy tool for practicing managers as well.

The book is organized into two parts. Operations research (OR) is covered in the first part: Chaps. 1–7. Chapter 1 discusses the evolution of OR and sets the stage for other chapters to follow. Chapter 2 introduces linear programming and formulation. Chapter 3 discusses the simplex method and a few special cases encountered in linear programming. Chapter 4 gives an insight into transportation algorithms and discusses the various types of assignment problems. Chapter 5 introduces network models and discusses two powerful statistical tools in project management, i.e., PERT and CPM. Chapter 6 covers game theory as a decision-making tool in business situations. Chapter 7 concludes the part on OR by introducing queuing models and presenting eight different models for evaluation. Chapter 1 consists of a number of problems designed to illustrate various concepts and models. Most of these problems are taken from examination papers of different universities and competitive examinations.

Engineering management is covered in the next part: Chaps 8–13. Chapter 8 introduces engineering management and summarizes its evolution and scope by looking at various management philosophies. Chapter 9 lays out the foundations of technology management through a comprehensive overview of strategic management with a technology flavor. Chapter 10 outlines the role of information technology in taking strategic decisions. Chapter 11 strives to emphasize the impact of R&D on strategy formulation and execution. Chapter 12 discusses the various aspects of managing projects and addresses the *how, why*, and *when* of project management. Chapter 13 discusses professional communications and lays out the guidelines for communicating better in a business context.

This book has been possible due to the efforts of Sagarika Ghosh (Sr. Editor with Springer), who appreciated the need for an easy-to-use and practical manual for engineering and management students, and who believed in my ability to deliver the same. My thanks also go to Dr. Srinidhi, who is currently associated with Hewlett Packard Company for his timely and constructive suggestions. He was instrumental in solving many of the problems in the book. Thanks are also due to Ms. Arpita Shetty, Project Assistant, IISc, who helped in keying in the material. I would also like to acknowledge PHI Learning Private Limited for permitting the use of some content from their publication, *Strategic Management—The Indian Context* (4th edition, 2012), authored by me. The staff of Springer has brought out this book in a very short time and in an attractive manner. My sincere thanks to the entire team.

But most of all, my thanks go out to my students and readers who have inspired the content and style of this book. I will consider my effort to be successful only if this book helps them understand the basics of the subject and advances their interest in the same.

Constructive suggestions for improvement of the book are welcome.

R. Srinivasan

# Contents

# About the Author

**R. Srinivasan** is Professor, Department of Management Studies, Indian Institute of Science, Bangalore, India. After his postgraduation in Engineering, he obtained his doctoral degree (Fellowship Management Program) from the Indian Institute of Management, Bangalore. Subsequently, he did his postdoctoral training from the University of Leeds, UK. He has more than 33 years of experience in both academia and industry, having served in some of India's reputed organizations like Tata Consulting Engineers (TCE), New Delhi, Administrative Staff College of India (ASCI), Hyderabad, and National Institute for Training in Industrial Engineering (NITIE), Mumbai. He serves as a visiting faculty and examiner in many reputed institutions in the country. His papers have been published in leading international and national journals and he has authored six books and has contributed to the Springer (2001) publication *Advertising Worldwide*. He is also a recipient of a number of awards, notably the International Statistical Institute Award (1983) and the Colombo Plan Award (1989). He has been invited to serve on the panel to develop International Technology Indicators by the Georgia Institute of Technology, Atlanta, USA, for 2002–2004, 2005–2006 and again for 2007–2008. His interest in the infrastructure development field has influenced his work on a number of assignments of societal and national interest like the socioeconomic impact analysis of the Bedthi hydel project and, more recently, an estimation of the demand for solar photovoltaic (SPV) cells in Gundlupet, a drought-prone area in Karnataka, India. The Indian Journal of Transport Management has rated the research article *Intermodal Choice in Passenger Transportation: Some Empirical Evidence* as the best research article in 2001. His research on *Dimensional Identification of International Positioning Process for Defence Systems* won the European Aeronautical Defence and Space Systems Award in 2009. He was invited in 2006 to share his thoughts and expertise in the European Masters Program on Management at Madrid, Spain. He is consulted regularly by well-known Indian national bodies including the Union Public Service Commission (UPSC), Indian Space Research Organisation (ISRO), and Ministry of Defence. His current research interests are developing strategic policy initiatives and strategic marketing.

# Part I
# Operations Research (OR)

# Chapter 1
# Introduction

The key takeaways for the reader from this chapter are listed below:

- How OR evolved and the origins of the subject
- What OR means?
- The implication of an OR model
- Solving the OR model and simulations.

## 1.1 Evolution of OR: Traced

Many of the concepts of operations research (OR) were used even before the actual term OR came into being like Queuing Theory, Inventory Control, and Statistical Quality Control. A simple economic order quantity (EOQ) model was developed by 1915 by Ford Harris which was later analyzed by R. H. Wilson in 1930. Around 1916, A. H. Erlang, a Spanish telephone engineer, developed many of the early theoretical developments of Queuing theory. Routine quality checks at Western Electric laboratory by Stewart resulted in design of control charts in 1924. Terms like consumers' risk, producers' risk, probability of acceptance, operations characteristic (OC) curve, lot tolerance percent defective (LTPD), and type I and type II errors were defined in 1925–1926. Sampling Inspection was introduced by Dodge in 1925.

World War II goaded scientists from different fields to research military operations for improving effectiveness with limited resources. This developed into an important aspect of OR methodology later on. Post-World War II saw OR methodologies being applied to maximize profitability with limited resources. Dantzig developed the simplex method in 1947 to solve linear programming (LP) problem. The Operations Research Society of America (ORSA) and the Institute of Management Science came into being in 1952 and 1953, respectively.

Development of high-speed digital computers has made it possible to apply some OR technologies to solve real-life problems. Probability theory and statistics have helped indeterminist situations.

R. Srinivasan, *Strategic Business Decisions*,
DOI: 10.1007/978-81-322-1901-9_1, © Springer India 2014

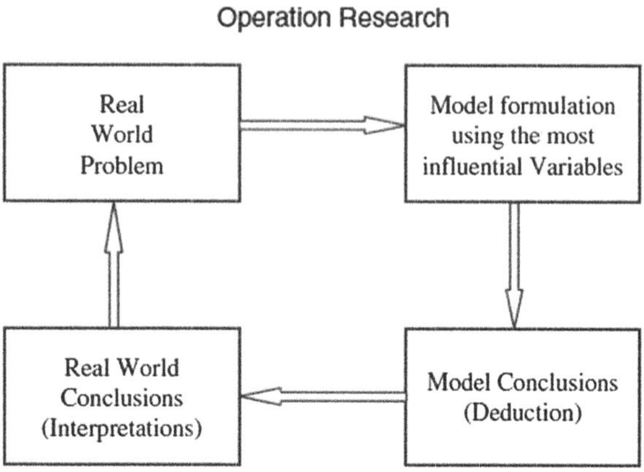


**Fig. 1.1** The steps in modeling any situation on hand

## 1.2 OR: Defined

According to ORSA, OR is a tool concerned with the design and operation of man–machine system scientifically, usually under conditions requiring the optimum allocation of limited resources. According to the Operations Society of Great Britain, OR is the application of scientific methods to compare problems arising in the direction and management of large systems of men, machines, materials and many in industry, business, and government.

## 1.3 Models

OR uses "models" to obtain solution of various realistic problems. Model represents abstraction of reality; examples include city road map to determine the shortest distance between an origin and destination, 3-D view of a factory to plan movement of materials, and regression equation to forecast demand for a product.

The steps in modeling any situation on hand are schematically represented in Fig. 1.1.

Step 1 involves identifying the management decisions with respect to the real-world problem.

Step 2 involves model formulation—for this, the parameters and the variables involved have to be identified, verbally defined, and giving symbols to each one of them; most influential variables are then related; these can be controllable and non-controllable; verbal relationships among the identified variables are stated using the known principles, data, and intuition.

Step 3 involves development of equations and statistical analysis to optimize certain performance measures and draw conclusions.

Step 4 involves the interpretation of model conclusions in the light of real-world problems' characteristics. These have to be tested and validated before implementation. The model is also to be reviewed as and when required.

The three main components of an OR model are alternatives, objective, and constraints.

The first critical step in any OR model is the definition of alternatives or decision variables of the problem. These are used to construct the objective function and constraints of the model.

The general format for an OR model is

Maximize/Minimize—**Objective Function**

Subject to

**Constraint 1** --------- (1)

**Constraint 2** --------- (2)

⋮

**Constraint n** --------- (N)

A **Feasible** solution satisfies all the above constraints. A Feasible solution becomes **Optimal** if it also yields the best (maximum or minimum) value of the objective function. If all the dominant alternatives for a particular problem situation are not identified, then it can result in a suboptimal solution. The "optimum" solution is best only for that model. If the model is a true representative of the real system, then the solution is optimal also for the real situation.

## 1.4 Solving the OR Model: Explained

There is no simple or general technique in OR that solves all mathematical models. The type and complexity of the mathematical model dictate the nature of the solution.

LP is the most prominent technique of OR. It involves strict linear objective and constraint functions. Integer programming (variables assume integer values), dynamic programming (original problem can be decomposed into smaller subproblems), network programming (problem is modeled as a network), and nonlinear programming (model functions are nonlinear) are some of the other OR techniques.

Solutions for most OR techniques are not obtained in formulas like closed forms. They are instead denoted by **algorithms**. An algorithm has fixed computational rules that are applied repetitively to the problem. Each repetition (called **iteration**) moves the solution closer to the optimum level. Since computations associated with each iteration are tedious and voluminous, algorithms are executed on the computer.

In some mathematical models, because of the complexity, none of the available optimization algorithms may solve it. In many cases, instead of seeking the *optimal* solution, a *good solution* using **heuristics** or *rule of thumb* can be thought of.

## 1.5 Queuing and Simulation Models

Queuing and simulation models are not optimization techniques; they study waiting times to determine measures of performance like average waiting time in the queue, average waiting time for service, and utilization of service facilities. Queuing makes use of probability and stochastic models to analyze waiting times. Simulation imitates the behavior of the real system to estimate the performance, whereas queuing models are purely mathematical, subject to specific assumptions that may limit their applications. Simulation models are flexible and can analyze any queuing situation. However, developing simulation models is costly with respect to both time and resources. Further, execution of simulation models even on the fastest computers is slow.

## 1.6 Phases of OR Study

OR in practice involves the following phases:

1. Problem definition
2. Model construction
3. Model solution
4. Model validation
5. Implementation.

Problem definition involves defining the problem scope. The investigation involves identifying three principal elements:

1. Decision alternatives
2. Determination of the objective function
3. Specification of limitations.

Model construction involves translation of the problem definition into mathematical relationships. If it fits into standard mathematical models such as LP, solution can be obtained using available algorithms. If the mathematical relationships turn out to be complex, one may have to simplify the model using heuristics, or simulation, as appropriate. Some models may require a combination of mathematics, heuristics, and simulation.

Model solution entails the use of well-defined optimization algorithms. Sensitivity analysis is done in this phase to study the behavior of the optimal solution when some parameters are changed. It is particularly useful when parameters cannot be estimated accurately.

Model validity checks whether or not the model predicts the behavior of the system under study adequately. The model should give sensible and acceptable solutions. For this, the model output can be compared with the historical output data; if under similar input conditions, the model reproduces past performance, and then, it is valid. However, past performance is not a guarantee of future performance. If no historical data are available, simultaneously many have to be used to verify the output of the mathematical model.

Implementation of solution involves translation of results into operating instructions, understood by individuals administering the recommended system. This is primarily the task of the "OR teams."

## 1.7 Review Questions

1. Trace the development of OR. How is "OR" now defined?
2. What is a "model"? What are the steps involved in an OR model?
3. What according to you is the approach to solve any OR problem?
4. In what aspects are queuing and simulation different from each other? Elaborate?
5. Describe the phases of an OR study?

# Chapter 2
# Introduction to Linear Programming

The key takeaways for the reader from this chapter are listed below:

- A good understanding of linear programming (LP) problems
- Formulation of the two-variable LP problem
- Understanding optimization in the contexts of minimization and maximization objective functions
- Representing a two-variable LP model graphically.

## 2.1 Introduction

LP is an optimization model in which the objective functions and constraints are strictly linear. It is used in a wide range of areas such as agriculture, transportation, economics, and industry. The advent of computers has made it the backbone of solution algorithms for other OR models including integer, stochastic, and non-linear programming.

In this chapter, we discuss a two-variable LP model and present its graphical solution. As we already know, any LP model will contain an objective function, set of constraints, and non-negativity restrictions.

Each of the components may involve one or more of the following:

- Decision variables
- Objective function coefficients
- Technical coefficients
- Resources availability.

## 2.2 Two-Variable LP Model

In this section, a graphical solution for a two-variable LP model is presented, though in reality there will be more than two variables in any situation. A typical product mix problem is considered as an example.

R. Srinivasan, *Strategic Business Decisions*,
DOI: 10.1007/978-81-322-1901-9_2, © Springer India 2014

### 2.2.1 Example: RM Company

**Illustration 2.1**

RM company produces two products, $P_1$ and $P_2$, from two raw materials, $R_1$ and $R_2$. Typical examples can be: Product $P_1$ produced from a milling machine using $R_1$ and $R_2$, and product $P_2$ produced from a drilling machine using the same raw materials $R_1$ and $R_2$. The basic data of the problem are given below (Table 2.1).

According to a market survey, the daily consumption of $P_1$ cannot exceed that of $P_2$ by more than 1 ton. Further, the maximum daily demand of $P_2$ is 2 tons.

It is required to determine the optimum (best) amounts of $P_1$ and $P_2$ that maximize the total daily profits.

**Solution**

Any OR model, including an LP model as mentioned in Chap. 1 has three basic components.

- Decision variable
- Objective (Goal)
- Constraints that need to be satisfied.

In the present case, the decision variables to be determined are the amounts of $P_1$ and $P_2$. We can define them as

$X_1$—Tons produced daily of $P_1$.
$X_2$ —Tons produced daily of $P_2$.

(*Note* A company would always like to maximize the profits and minimize the costs. Hence, the objective function would be *Maximize* in case of profits and *Minimize* in case of costs.)

According to the problem, the company wants to maximize its profits. Hence, the objective function can be formulated as

$$\text{Maximize } Z = 5X_1 + 4X_2$$

The constraints are with respect to raw material and demand. The raw material constraint can be verbally expressed as

$$\begin{pmatrix}\text{Usage of raw material}\\ \text{by both } P_1 \text{ and } P_2\end{pmatrix} \leq \begin{pmatrix}\text{maximum of raw material}\\ \text{availability}\end{pmatrix}$$

For the present problem,

$$\text{Usage of raw material } R_1/\text{day} = 3x_1 + 2x_2 \text{ tons}$$
$$\text{Usage of raw material } R_2/\text{day} = 1x_1 + 2x_2 \text{ tons}$$

**Table 2.1** The basic data of the problem

| | Tons of raw material per ton of | | Max daily availability (tons) |
|---|---|---|---|
| | $P_1$ | $P_2$ | |
| Raw material $R_1$ | 3 | 2 | 12 |
| Raw material $R_2$ | 1 | 2 | 6 |
| Profit/ton (Rs. 000) | 5 | 4 | |

Daily availability of $R_1$ and $R_2$ is limited to 12 and 6 tons. Hence, the constraints can be expressed as

$$3X_1 + 2X_2 \leq 12(R/M\ R_1) \tag{2.1}$$

$$1X_1 + 2X_2 \leq 6(R/M\ R_2) \tag{2.2}$$

The demand constraints of daily demand of $P_1$ cannot exceed that of $P_2$ by more than 1 ton, which translates to

$$X_2 - X_1 \leq 1 \tag{2.3}$$

Similarly, the second demand constraint of the maximum daily demand of $P_2$ is 2 tons, which translates to

$$X_2 \leq 2 \tag{2.4}$$

$X_1$ and $X_2$ cannot assume negative values. The nonnegativity constraints, therefore, are $x_1$, $x_2 \geq 0$ and the complete RM company model can be written as

$$\begin{aligned} \text{Maximize } Z = {} & 5X_1 + \ 4X_2 \\ \text{subject to (s.t) } & 3X_1 + 2X_2 \leq 12 \\ & 1X_1 + 2X_2 \leq 6 \\ & X_2 - X_1 \leq 1 \\ & X_2 \leq 2 \\ & X_1, X_2 \geq 0 \end{aligned}$$

Any values of $X_1$ and $X_2$ that satisfy all the above constraints constitute a feasible solution. For example: $X_1 = 2$ and $X_2 = 1$ is a feasible solution, since all the constraints are satisfied, including the nonnegativity constraints.

The optimum feasible solution yields maximum total profit while satisfying all the constraints. There can be a number of feasible solutions to a problem but finding the optimum solution requires following a systematic procedure, which will be explained in Chap. 3. The next solution gives the optimum solution obtained using the graphical method. In the above example, the objective function and constraints are all linear. Linearity means any LP must satisfy two properties: proportionality and additivity.

Proportionality requires that the contribution of each decision variable in the objective function and its requirements in the constraints to be directly proportional to the variable. In the RM Company model, $5X_1$ and $4X_2$ give the profits for producing $X_1$ and $X_2$ tons of $P_1$ and $P_2$, respectively, with unit profits per ton, 5 and 4, providing the constants of proportionality. Suppose RM Company gives quantity discounts when sales exceed certain amounts, then profit will no longer be proportional to $X_1$ and $X_2$.

Additivity requires that the total contribution of all the variables in the objective function and their requirements in the constraints are the direct sum of the individual contributions or requirements of each variable. In the RM Company model, the total profit equals the sum of the individual profit components. Suppose the two products compete for the same market share such that increase in sales of one adversely affects the other, then additivity is not satisfied.

**Illustration 2.2**

For the RM Company Model, construct each of the following constraints and express them with a constant right-hand side:

(a) The daily demand for $P_2$ exceeds that of $P_1$ by at least 1 ton.
(b) The daily usage of raw material is at most 6 tons and at least 3 tons.
(c) The demand for $P_1$ cannot be less than the demand for $P_2$.
(d) The minimum quantity that should be produced of both the paints $P_1$ and $P_2$ is 3 tons.
(e) The proportion of interior paint to the total production of both the paints $P_1$ and $P_2$ must not exceed 5.

**Solution**

All the formulations are given below:

(a) Here, the daily demand for $P_2$ exceeds that of $P_1$ at least by 1 ton, so we have the formulation as

$$X_2 - X_1 \geq 1 \text{ or } -X_1 + X_2 \geq 1$$

(b) The daily usage of raw material is at most 6 tons $-X_1 + X_2 \leq 6$. The daily usage of raw material is at least 3 tons $-X_1 + 2X_2 \geq 3$.

(c) If the demand for $P_1$ cannot be less than $P_2$, we have the formulation as $X_2 - X_1 = 0$ or $X_2 = X_1$.

(d) Here, it is given that the minimum quantity to be produced is 3 tons. We can formulate this as $X_1 - X_2 \geq 3$.

(e) The proportion can be formulated as $X_2/X_1 + X_2 = 0.5$ or $0.5X_1 - 0.5X_2 \geq 0$.

**Illustration 2.3**
Suppose it is asked to determine the best "feasible" solution.

Determine the best feasible solution among the following (feasible and infeasible) solutions of the RM Company model:

(a) $X_1 = 1, X_2 = 4$
(b) $X_1 = 2, X_2 = 2$
(c) $X_1 = 3, X_2 = 1.5$
(d) $X_1 = 2, X_2 = 1$
(e) $X_1 = 2, X_2 = -1$.

**Solution**
In order to find the best feasible solution, we have to substitute all the above values of $X_1$ and $X_2$ in all the constraint equations of the RM Company Model. The obtained values will have to be less than the right-hand side values of the constraint equations. Then, the solution will be graded as feasible.

(a)

$$X_1 = 1, \ X_2 = 4$$

Substituting the above values in the constraint equations, we get

$$6.1 + 4.4 = 22(<24)$$
$$1.1 + 2.4 = 9( > 6)$$

We can see above that the rule for feasibility is violated in the second constraint as the obtained value is more than the right-hand side value of the equation. Hence, this is infeasible.
(b)

$$X_1 = 2, \ X_2 = 2$$

Substituting the above values in the constraint equations, we get

$$6.2 + 4.2 = 20(<24)$$
$$1.2 + 2.2 = 6(=6)$$
$$-1.2 + 1.2 = 0(<1)$$
$$1.2 = 2(=2)$$

Here, we see that all feasibility rules are satisfied for all the constraint equations. Hence, this is a feasible solution.

The objective function value will be $Z = 5.2 + 4.2 = 18$.

(c)

$$X_1 = 3,\ X_2 = 1.5$$

Substituting the above values in the constraint equations, we get

$$\begin{aligned} 6.3 + 4.1.5 &= 24(=24) \\ 1.3 + 2.1.5 &= 6(=6) \\ -1.3 + 1.1.5 &= -1.5(<1) \\ 1.1.5 &= 1.5(<2) \end{aligned}$$

Here, we again see that all feasibility rules are satisfied for all the constraint equations. Hence, this is also a feasible solution.

The objective function value will be $Z = 5.3 + 4.1.5 = 21$.

(d)

$$X_1 = 2,\ X_2 = 1$$

Substituting the above values in the constraint equations, we get

$$\begin{aligned} 6.2 + 4.1 &= 16(<24) \\ 1.2 + 2.1 &= 4(<6) \\ -1.2 + 1.1 &= -1(<1) \\ 1.1 &= 1(<2) \end{aligned}$$

Here, again all feasibility rules are satisfied for all the constraint equations. Hence, this is also a feasible solution.

The objective function value will be $Z = 5.2 + 4.1 = 14$.

(e)

$$X_1 = 2,\ X_2 = -1$$

Here, we see that the value of $X_2 = -1$, i.e., negative. So, we can straightaway say that this is an infeasible solution

From all the above calculations, we conclude that option *c* is the best "feasible" solution as the value of the objective function is maximum, i.e., maximum profits are obtained by having $X_1 = 3$, and $X_2 = 1.5$.

**Illustration 2.4**

A state government owned farm measuring about 150 acres sells all the ladies fingers, radishes, and onions it can produce. The price obtained is Rs. 1.5/kg for ladies fingers and onions, and Rs. 3/kg for radishes. Wages are paid to the farmers at Rs. 25 per man-day. It is required for the farmers to put in 4 man-days for ladies fingers, 5 man-days for onions, 6 man-days for radishes, and a total of 380 man-days of effort to be put in for the entire production. Fertilizers required for the same are available at Rs. 0.60/kg, and the quantity required per acre is 110 kgs for ladies fingers and onions, and 75 kgs for radishes. It is also required that the average yield per acre should be 2,500 kgs for ladies fingers, 1,500 kgs for onions, and 2,000 kgs for radishes. Formulate the above as an LP so as to give the state government maximum profits.

**Solution**

The above problem gives many details, but the details to be extracted for the purpose of formulation are average yield per acre, measurement of the land, man-days needed for production, cost of labor, and any other costs, if specified.

$$\text{Profit} = \text{Total Sales Value} - \text{Total Cost}$$

Total cost here accounts for labor cost and fertilizer cost.

$$\text{Hence } [\text{Profit} = \text{Total Sales Value} - \text{Total Cost (Labor + Fertilizer)}]$$

Now, the equations can be arrived at.

$$\begin{aligned} \text{For ladies fingers, profit} &= 2,500 - (25.4 + 110.0.6) = 2,334 \\ \text{For onions, profit} &= 1,500 - (25.5 + 110.0.6) = 1,309 \\ \text{For radishes, profit} &= 2,000 - (25.6 + 75.0.6) = 2,805 \end{aligned}$$

Hence, the objective function becomes

$$\begin{aligned} \text{maximize} &= 2334X_1 + 1309X_2 + 2805X_3 \\ \text{subject to} \quad & X_1 + X_2 + X_3 \leq 150 \text{ (Land constraint)} \\ & 4X_1 + 5X_2 + 6X_3 \leq 380 \text{ (Labor constraint)} \\ & X_1, X_2, X_3 \geq 0 \end{aligned}$$

**Illustration 2.5**

A soft drink plant has two bottling machines, *Y* and *Z*. *Y* is designed for 8-ounce bottles and *Z* for 16-ounce bottles. However, each can be used on both types with some loss in efficiency. The following data are made available (Table 2.2).

The machine can be run for 8 h a day, and 5 days a week. The profit is 15 and 25 paise on 8-ounce and 16-ounce bottles, respectively. The plant has to limit the

**Table 2.2** Data obtained by the two bottling machines, *Y* and *Z*

| | 8-ounce bottles | 16-ounce bottles |
|---|---|---|
| *Y* | 100/min | 40/min |
| *Z* | 60/min | 75/min |

weekly production to 3,00,000 ounces, and the market can absorb 25,000 8-ounce and 7,000 16-ounce bottles per week. Formulate the above problem as an LP so as to maximize the plants profits subject to all production and marketing restrictions.

**Solution**

Let $X_1$ units of 8-ounce bottles and $X_2$ units of 16-ounce bottles be produced. As the profits are known, the objective functions can be directly arrived at. The total profit of the plant is given by $0.15X_1 + 0.25X_2$.

The objective function will be maximize d as $Z = 0.15X_1 + 0.25X_2$. Since machines *Y* and *Z* work for 8 h a day and 5 days a week, the total working time for machines *Y* and *Z* will become 2,400 min per week. Therefore, the time constraints are

$$X_1/100 + X_2/40 \leq 2,400 \text{ (for machine } Y)$$
$$X_1/60 + X_2/75 \leq 2,400 \text{ (for machine } Z)$$

The other constraints are

$$8X_1 + 16X_2 = 3,00,000 \text{ (Production constraint)}$$
$$\left.\begin{array}{l} 0 \leq X_1 \leq 25,000 \\ 0 \leq X_2 \leq 7,000 \end{array}\right\} \text{(Marketing constraints)}$$
$$X_1, X_2 \geq 0$$

**Illustration 2.6**

The quality control department of an organization has recruited 8 and 10 inspectors in grades 1 and 2, respectively, for performing quality inspections on products. It is imperative for the inspectors to maintain 95 % accuracy in their work. In order to achieve this, the inspectors can check not more than 20 and 15 pieces, respectively, in an hour, and not more than 1,450 pieces in a day. Labor costs account to Rs. 4/h for grade 1 inspectors, and Rs. 5/h for grade 2 inspectors. If overloaded, there are possibilities for inspection errors and this costs Rs. 3.

It is required to find the optimal assignment of inspectors that minimizes the daily costs for the organization.

**Solution**
The objective of this problem is to minimize the costs. Hence, the objective function has to be of minimization type. There is 5 % chance of error in the quality control inspections. Hence, the costs incurred by the organization will be:

$$\text{For grade 1 inspectors: } 4 + (0.05.3.18) = \text{Rs.}\,6.7/\text{h} = 6.7.8 = 52.6/\text{day}$$
$$\text{For grade 2 inspectors: } 5 + (0.05.3.14) = \text{Rs.}\,7.1/\text{h} = 7.1.8 = 56.8/\text{day}$$

Now, we can formulate the objective function.

$$\begin{aligned} \text{Minimize}\, Z &= 52.6X_1 + 56.8X_2 \\ \text{Subject to}\quad & X_1 \le 8, X_2 \le 10 (\text{Recruitment constraint}) \\ & 144X_1(18.8) + 112X_2(14.8) \ge 1,450 (\text{Capability constraint}) \\ & X_1, X_2 \ge 0 \end{aligned}$$

## 2.3 Graphical Solution to an LP Model

The graphical approach to solve an LP problem obtains a feasible solution to the problem, provided the number of variables involved is two. This section demonstrates the graphical approach under possible conditions.

### *2.3.1 Graphical Solution of a Maximization Model*

**Illustration 2.7**
Solve the following LP problem by the graphical method.

$$\begin{aligned} \text{Maximize}\, Z &= 5X_1 + 12X_2 \\ \text{Subject to}\quad & 10X_1 + 20X_2 \le 100 \\ & 8X_1 + 7X_2 \le 56 \\ & X_1, X_2 \ge 0 \end{aligned}$$

Step 1: We have to first account for nonnegativity constraints $X_1$, $X_2 = 0$. In Fig. 2.1, $X_1$ and $X_2$ on the horizontal and vertical axes, respectively, represent $P_1$ and $P_2$. It follows that the nonnegativity constraints restrict the solution space to the first quadrant.

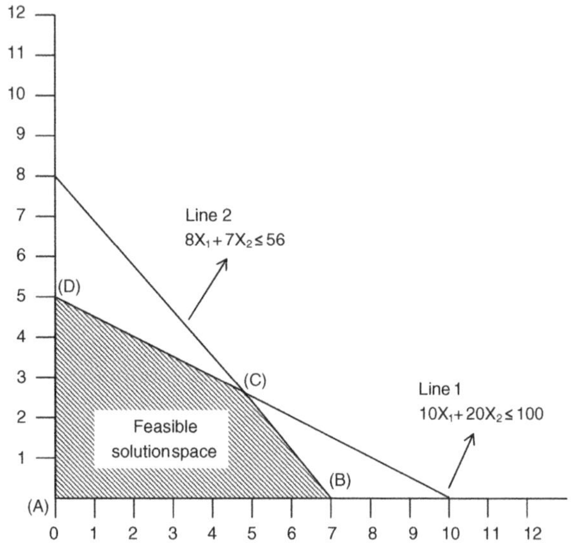


**Fig. 2.1** Nonnegativity constraints

To account for the remaining constraints, each inequality is replaced with equations.

Locating two distinct points on it draws a graph of the resulting straight line. In the example, if we replace $10X_1 + 20X_2 = 100$ with the straight line $10X_1 + 20X_2 = 100$, then two points can be determined by setting $X_1 = 0$ and $X_2 = 0$. By setting $X_1 = 0$, we get $20X_2 = 100$, i.e., $X_2 = 5$. Similarly by setting $X_2 = 0$, we get $10X_1 = 100$, i.e., $X_1 = 10$. Thus, the line passes through (10, 0) and (0, 5), which is shown as line 1. Line 1 represents the first constraint. Similarly for the second constraint, we get $X_1 = 7$ by setting $X_2 = 0$, and $X_2 = 7$ by setting $X_1 = 0$. This line passes through (7, 0) and (0, 8) as shown in Line 2. Line 2 represents the second constraint. $X_1$ and $X_2 = 0$ eliminate the second, third, and fourth quadrants of the $X_1$ and $X_2$ plane.

Step 2: *Optimum Solution* The closed space $A$–$B$–$C$–$D$ represents the feasible solution space. The objective function value at each of the corner points is computed by substituting its coordinates in the objective function as:

$$Z(A) = 5(0) + 12(0) = 0$$
$$Z(B) = 5(10) + 12(0) = 50$$
$$Z(C) = 5(5) + 12(3) = 61$$
$$Z(D) = 5(2) + 12(5) = 60$$

Since the objective function here is maximization, the solution corresponding to the maximum $Z$ value is to be selected as the optimum solution. The $Z$ value is maximum for the corner point $C$. The corresponding solution is $x_1 = 5$ and $x_2 = 3$,

**Fig. 2.2** Feasible region

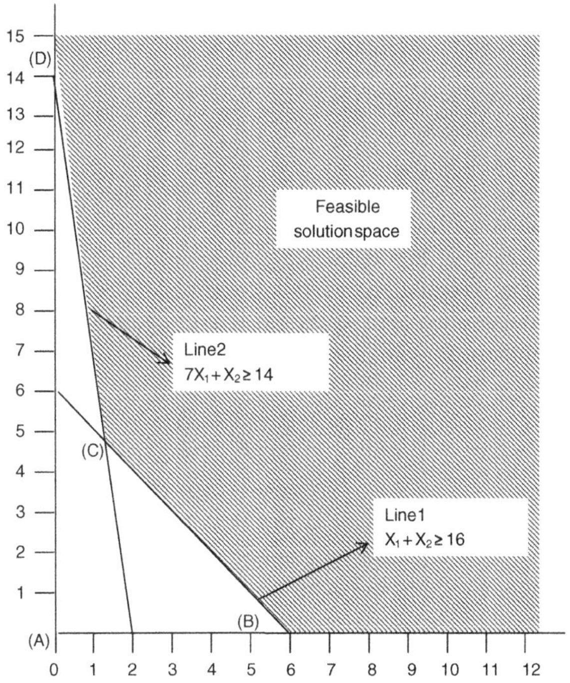


and $Z$ (Optimum) $= 61$. This calls for a daily mix of 5 tons of $P_1$ and 3 tons of $P_2$, giving an associated daily profit of Rs. 61,000.

## *2.3.2 Solution for a Minimization Problem*

**Illustration 2.8**

Solve the following problem by graphical method:

$$\begin{aligned} \text{Minimize } Z &= 2X_1 + 3X_2 \\ \text{subject to } \quad & X_1 + X_2 \geq 6 \\ & 7X_1 + X_2 \geq 14 \\ & X_1 X_2 \geq 0 \end{aligned}$$

**Solution**
Let us compute the coordinates on the *XY* plane relating to constraint equations.

First constraint gives us $X_1 = 6$, when $X_2 = 0$, and $X_2 = 6$, when $X_1 = 0$, represented by
Line 1 in Fig. 2.2.

Second constraint gives us $X_1 = 2$, when $X_2 = 0$, and $X_2 = 14$, when $X_1 = 0$, represented by
Line 2 in Fig. 2.2.

The feasible region is represented by *F–D–C–B–G*. Now, it is required to calculate the value of the objective function at these corner points to obtain the optimum solution.

$$Z(D) = 2.0 + 3.14 = 42$$
$$Z(C) = 2.4/3 + 3.14/3 = 50/3$$
$$Z(D) = 2.6 + 3.0 = 12$$

As our objective function is of minimization type, it is required to choose the minimum value from the above values, which correspond to corner point *D* of value 12. Therefore, the optimum solution is:

$$X_1 = 6,\ X_2 = 0,\ \text{and } Z\ (\text{optimum}) = 12.$$

## 2.4 Review Questions

1. Explain the concepts of an LP model?
2. What are the assumptions of an LP problem?
3. Elaborate the steps involved in solving an LP problem graphically?
4. A textile company can use any or all of the three different processes for weaving its standard white polyester fabric. Each of these production processes has a weaving machine setup cost and per-square-meter processing cost. The costs and capacities of the three production processes are as given below.

| Process number | Weaving machine setup cost (Rs.) | Processing cost/$m^2$ (Rs.) | Maximum daily capacity ($m^2$) |
|---|---|---|---|
| 1 | 75 | 15 | 1,000 |
| 2 | 120 | 10 | 1,500 |
| 3 | 150 | 8 | 1,750 |

The daily demand forecast for its white polyester fabric is 4,000 $m^2$. The company's production manager wants to make a decision concerning which combination of production process is to be utilized to meet the daily demand forecast and at what production level of each selected production process to be operated to minimize total production costs.

5. Consider the cargo-loading problem, where five items are to be loaded on a vessel. The weight ($w_i$) and the volume ($v_i$) of each unit of the different items as well as their corresponding returns per unit ($r_i$) are as shown below.
The maximum cargo weight (W) and volume (V) are given as 112 and 109, respectively. It is required to determine the optimal cargo load in discrete units of each item such that the total return is maximized. Formulate the problem as an integer programming model. (*Hint*: Similar to LP formulation)

| Item $i$ | $w_i$ | $v_i$ | $r_i$ |
|---|---|---|---|
| 1 | 4 | 1 | 5 |
| 2 | 7 | 8 | 8 |
| 3 | 2 | 6 | 7 |
| 4 | 1 | 5 | 6 |
| 5 | 6 | 4 | 5 |

6. A firm manufactures two headache pills *A* and *B*. Size A contains 2 grains of aspirin and 5 grains of codeine. Size B contains 1 grain of aspirin, 9 grains of bicarbonate, and 6 grains of codeine. It is found by customers that it requires at least 12 grains of aspirin, 74 grains of bicarbonate, and 24 grains of codeine for providing immediate relief. Formulate the problem as an LPP. (*Hint* Minimization problem $-\min\ X_1 + X_2$, s.t. $2X_1 + X_2 = 12$, $X_1 + 6X_2 = 24$, $5X_1 + 8X_2 = 24$; $X_1, X_2 = 0$).
7. A manufacturer has three machines *A*, *B*, and *C* which produce three different articles *P*, *Q*, and *R*. The different machine times required per article, the amount of time available in any week on each machine, and the estimated profits per article are tabulated below.

| Article | Machine time (in hrs) | | | Profit per article (in Rs.) |
|---|---|---|---|---|
| | *A* | *B* | *C* | |
| *P* | 8 | 4 | 2 | 20 |
| *Q* | 2 | 3 | 0 | 6 |
| *R* | 3 | 0 | 1 | 8 |
| Available machine (h) | 250 | 150 | 50 | |

Formulate the problem as an LPP. (*Hint*: Similar to illustration).

# Chapter 3
# The Simplex Method

The key takeaways for the reader from this chapter are as follows:

- Introduces the simplex method for solving LP
- Gives iterative procedure and graphical solution
- Discusses Big M method, duality, and dual simplex
- Special cases.

## 3.1 Introduction

There are many algorithmic methods to solve LP problems. The graphical method forms the basis for development of algebraic simplex method. Transition from graphical to algebraic solution is shown in Fig. 3.1.

## 3.2 Simplex Method

Simplex method is a systematic algorithm moving from one basic feasible solution to another so that the objective function value is improved. This procedure of jumping from vertex to vertex is repeated. If the objective function improves at each jump, there is no need to go back to the vertex already covered. As the number of vertices is finite, the process leads to the optimal vertex in a finite number of steps.

Simplex algorithm is a step-by-step (iterative) procedure for solving LP problems. It involves:

- Having initial basic solution to constraint equations.
- Testing whether it is an optimal solution.
- Improving the first initial solution by a set of roles and repeating the process till an optimal solution is obtained.

R. Srinivasan, *Strategic Business Decisions*,
DOI: 10.1007/978-81-322-1901-9_3, 

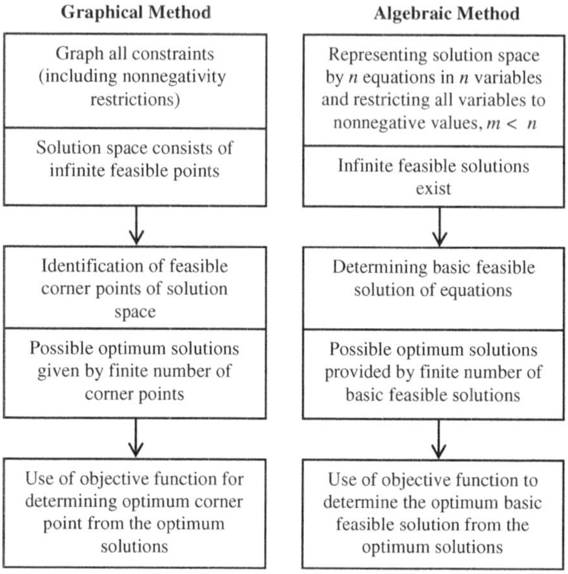


**Fig. 3.1** Transition from graphical to algebraic solution

The LP solution space is made under two conclusions:

1. All the constraints except nonnegative restrictions are equations with positive (nonnegative) equations.
2. All the variables are nonnegative (positive).

This procedure requires at most m (equals the number of equations) nonzero variables at each step. If it becomes less than m nonzero variables at each step, then degeneracy sets in. A feasible solution at any iteration is related to the feasible solution at the successive iterations as listed below:

One of the non-basic variables (which are 0 (zero) now) at one iteration becomes basic variable (BV) in the following iteration. This variable is called *entering variable*. One of the basic variables (which are nonzero now) at one iteration becomes non-basic (zero) in the following iteration. This variable is called *departing variable*. Other non-basic variables remain zero, and the other basic variables, in general, remain nonzero.

Suppose, we state an LPP in the standard form as follows:

$$\text{Maximize } Z = C_1 X_1 + C_2 X_2 + \ldots\ldots + C_n X_n + 0X_{n+1} + 0X_{n+2} + \ldots\ldots + 0X_{n+m} \ldots\ldots \quad (1)$$

subject to (s.t)

$$\left.\begin{array}{l} A_{11} X_1 + A_{12} X_2 + \ldots\ldots + A_{1n} X_n + X_{n+1} = B_1 \\ A_{11} X_1 + A_{12} X_2 + \ldots\ldots + A_{1n} X_n + X_{n+1} = B_1 \\ \ldots\ldots \\ \ldots\ldots \\ A_{m1} X_1 + A_{m2} X_2 + \ldots\ldots + A_{mn} X_n + \ldots\ldots + X_{n+m} = B_m \end{array}\right\} \ldots\ldots\ldots\ldots \quad (2)$$

$$X_1 \geq 0, X_2 \geq 0 \ldots\ldots X_n \geq 0, X_{n+1} \geq 0 \ldots\ldots X_{n+m} \geq 0$$

An obvious basic feasible solution of $m$ equations (2 above) is taken as: $X_1 = X_2 = \ldots X_n = 0$; $X_{n+1} = B_1$; $X_{n+2} = B_2$; $\ldots X_{n+m} = B_n$. The value of the objective function is zero for solution $X_1, X_2, X_3, \ldots X_n$ (each equal to zero) that are *non-basic variables* and the remaining $(X_{n+1}, X_{n+2}, \ldots X_{n+m})$ are basic variables.

Simplex method that forms the basic building block for all methods is based on the concept of solving simultaneous equations.

### *3.2.1 Computational Procedure of Simplex Method*

The iterative simplex procedure involves the following steps.

Step 1. The problem should be converted to maximization type if it is of minimization type by multiplying both sides of the objective function by $-1$.

If the objective function is Min $Z = X_1 - 3X_2 + 4X_3 \ldots - X_n$, then it should be rewritten as

$$\text{Maximize } Z' = -X_1 + 3X_2 - 4X_3 \cdots + X_n, \quad \text{where } Z' = -Z.$$

Step 2. The constraint values of the constraint equations ($B_i$'s) should be $\geq 0$. If any of the $B_i$'s is $\leq 0$, multiply the corresponding constraints on both sides of the constraint equation by $-1$, such that all $B_i$'s are $\geq 0$.

Step 3. All the inequalities should be changed to equations by adding slack or surplus variables, if required.

Step 4. Artificial variables should be added to those constraint equations with $(\geq)$ or $(=)$ signs, such that an identity basis matrix is obtained. (Artificial variables are introduced in Big $M$ method)

Step 5. Using the data available in the standard LPP, construct the starting/initial simplex table. This table provides the initial basic feasible solution (IBFS).

| Basic Variable (B.V.) | $C_B X_B$ | $X_1X_2\ldots S_1S_2\ldots A_1\ldots$ | Minimum Ratio $(X_B/X_k)$ |
|---|---|---|---|
| | | | |
| | $Z_j = \sum(C_B X_B)$ | $\Delta_1\Delta_2\ldots\Delta S_1\ldots\Delta A_1$ | $\leftarrow \Delta_j$ |

(*Note* In the IBFS, the slack or surplus variables $(S_1, S_2\ldots)$ occupy the place of BV. In cases wherein both artificial variables and surplus variables are introduced to balance an equation (with $\geq$ signs), such as in Big $M$ method, the respective artificial variables will have to be used as basic variables in place of surplus variables. These will be replaced by the decision variables $(X_1, X_2\ldots)$ in successive iterations.)

Step 6. Obtain the values of $\Delta_j$ by using the formula.

$$\Delta_j = Z_j - C_j \quad \text{or} \quad \Delta_j = C_B X_j - C_j. \tag{3.1}$$

The values thus obtained will have to be examined. This can result in three different possibilities.

(a) All values of $\Delta_j > 0$, in which case the basic feasible solution is optimal.
(b) Some values in $\Delta_j$ row are $<0$, and all $X_{rj} \leq 0$ for at least one of the corresponding $X_j$. The solution will be unbounded in this case.
(c) Some $\Delta_j \leq 0$, and all corresponding $X_j'$ s have at least one $X_{ij} > 0$. In this case, further improvement of solution is possible.

Step 7. By replacing one of the variables present in the basis matrix by the one outside the basis matrix, further improvement is achieved. Such a variable can be selected by adopting the following rules:

(a) **Selection of Incoming Variable**
The value of $k$ for which $\Delta_k = \min \Delta_j$, then $X_k$ will be the variable entering the basis matrix, or in other words $X_k$ will be the incoming variable.
(b) **Selection of Outgoing Variable**
This variable is the departing variable and will be $\beta_r$, and $r$ will be determined by the minimum ratio rule.

$$\frac{X_{Bh}}{X_{rk}} = \min \left[ \frac{X_{Bi}}{X_{ik}} \right], \quad \text{where } X_{ik} > 0 \tag{3.2}$$

Step 8. Construct further improvement simplex tables.
Step 9. Go to Step 6, then to 8 and 9, if required. Repeat this process till optimality is reached.

Let us take an example to illustrate the simplex algorithm

**Illustration 3.1**
**Consider the LP**

$$\begin{aligned} \text{Maximize } Z = {} & 6X_1 + 8X_2 \\ \text{subject to } \quad & 4X_1 + 4X_2 \leq 40 \\ & 5X_1 + 10X_2 \leq 60 \\ & X_1, X_2 \geq 0 \end{aligned}$$

**Solution**
The standard form of the above LP is given below.

$$\begin{aligned} \text{Maximize } Z &= 6X_1 + 8X_2 + 0S_1 + 0S_2 \\ \text{subject to } \quad & 4X_1 + 4X_2 + S_1 = 40 \\ & 5X_1 + 10X_2 + S_2 = 60 \\ & X_1, X_2, S_1, S_2 \geq 0 \end{aligned}$$

where $S_1$ and $S_2$ are slack variables, introduced to balance the constraints.

[*Note* There are two types of variables that can be introduced to balance the constraints—slack and surplus. Slack variables are introduced when the inequality is of the type $\leq$, and surplus variables are introduced when the inequality is of the type $\geq$. When these variables are introduced, the inequalities change into equalities, thus balancing the equation. (*Surplus variables will be dealt with at a later stage.*)]

Canonical represents the form in which each constraint has a BV. As mentioned earlier, a BV has unit coefficient in one of the constraints and zero in the remaining constraints. If all the constraints are of $\leq$ type, then the standard form is treated as canonical form.

## Solution

**Simplex Table 1**

| | | $C_j \rightarrow$ | 6 | 8 | 0 | 0 | |
|---|---|---|---|---|---|---|---|
| Basic Variable (B.V.) | $C_B$ | $X_B$ | $X_1$ | $X_2$ | $S_1$ | $S_2$ | Min ratio ($X_B/X_K$) |
| $S_1$ | 0 | 40 | 4 | 4 | 1 | 0 | 40/4 = 10 |
| $S_2$ | 0 | 60 | 5 | [10] | 0 | 1 | 60/10 = 6← |
| | $Z_j = 0$ | | - 6 | - 8 | 0 | 0 | ← $\Delta_j$ |
| | | | | ↑ Min $\Delta_j$ | | | |

1. The initial column represents the basic variables. Here, $S_1$ and $S_2$ are the basic variables, and hence, we write down under the first column.
2. The values under $C_B$ column always remain zero in the IBFS table.
3. The right-hand values in the construction equations, i.e., 40 and 60, should be taken down under the $X_B$ column.
4. The $C_j$ row values (written on top of decision variables) in the IBFS are the coefficients of the decision variables ($X_1$, $X_2$), and BV ($S_1$, $S_2$) in the objective function of standard form of the LP. This remains the same throughout the calculations.
5. The rest of the table is filled with the respective coefficient values from constraint equations as above.
6. The value of $Z_j = \sum C_B X_B = (0 \times 40) + (0 \times 60) = 0$.
7. For $\Delta_j$ row

$$\Delta_1 = C_B X_1 - C_j = (0 \times 4) - 6 = -6$$
$$\Delta_2 = C_B X_2 - C_j = (0 \times 4) - 8 = -8$$
$$\Delta_3 = C_B X_3 - C_j = (0 \times 1) - 0 = 0$$
$$\Delta_4 = C_B X_4 - C_j = (0 \times 0) - 0 = 0$$

(a) **To find out entering element/variable**
From the values obtained in the $\Delta_j$ row, identify the most negative value. The most negative value is $(-8)$ and mark that respective column with (↑). This is the entering variable column, i.e., $X_2$ is the entering variable which could replace either $S_1$ or $S_2$.

(b) **To find out leaving element/variable**
Find out ratio values $(X_B/X_k)$ under the leaving variable column and enter these values under the minimum ratio column.

$$\text{For} \quad S_1,\ X_B/X_k = 40/4 = 10$$
$$\text{For} \quad S_2,\ X_B/X_k = 60/10 = 6$$

From the calculated ratios, identify the least value. Here, the value is (6) and mark this row with (←). This implies $S_2$ is the leaving element, i.e., $X_2$ replaces $S_2$.

8. **Identification of Pivot Element**
After the leaving and entering variables are identified, the element at the intersection of the leaving variable row and entering variable column is the pivot element as shown in the table. Here, 10 is the element at the intersection, and hence is the pivot element.
Now, proceed to the next iteration, where $X_2$ replaces $S_2$.

**Simplex Table 2**

| | | $C_j$ → | 6 | 8 | 0 | 0 | |
|---|---|---|---|---|---|---|---|
| Basic Variable (B.V.) | $C_B$ | $X_B$ | $X_1$ | $X_2$ | $S_1$ | $S_2$ | Min ratio $(X_B/X_K)$ |
| $S_1$ | 0 | 16 | 2 | 0 | 1 | - 2/5 | 16/2 = 8 ← |
| $X_2$ | 8 | 6 | 1/2 | 1 | 0 | 1/10 | $\frac{6}{1/2} = 12$ |
| | $Z_j = 48$ | | - 2 | 0 | 0 | 8/10 | |
| | | | ↑ | | | | |

Notice that $X_2$ has replaced $S_2$. Here, the $C_B$ value for $X_2$ row will be 8, which must be taken from the $C_j$ row of the corresponding entering variable.

9. **Simplex Table Calculations**
In this book, we propose a simple method to calculate the values in successive iterations, which uses the formula

New element = Old element
− (Its entering column coefficient × New pivot element)
(3.3)

Initially, after identification of pivot element, divide all the values in the respective row by the pivot element. These are the values for the entering variable row in the next immediate iteration.
Here, 10 is the pivot element in simplex Table 1. Dividing all elements in the $S_2$ row of IBFS by 10 gives the values for $X_2$ row in the next immediate iteration.
Now, use formula (3.3) to find the values for $S_1$ row.
The new element is the new value which will be obtained from our calculations.
The old element is the already existing value in the previous iteration.
The entering column coefficients will always be the values in the pivot column corresponding to the respective row. Here, values for $S_1$ row are to be calculated. The value in the $S_1$ row above the pivot element 10 is (4). This will be the entering column coefficient for computing values in the entire $S_1$ row. These values have to be identified if there are more rows in the same way, and this remains fixed for that respective row only. This value is to be taken from the previous iteration.
The new pivot elements keep changing for every column. But these elements have to be chosen from the entering variable row after filling up the values for the same in the next immediate iteration.
Here, $X_2$ row values are filled up by dividing $S_2$ row values by pivot element (10). These values are the new pivot elements for the respective columns.

6 is the new pivotal element for $X_B$ column.
1/2 is the new pivotal element for $X_1$ column.
0 is the new pivotal element for $S_1$ column.
0/10 is the new pivotal element for $S_2$ column.

(*Note* The values in the pivotal column expect the pivot element remain zeros. Formula 3.1 does not apply to this column.)
From formula 3.1, compute the values for $S_1$ row.

(a) Under $X_B$ column:

Old element = 40
Entering column coefficient = 4 (look in IBFS)
New pivot element = 6 (value in $X_2$ row in the next iteration)
Therefore, value = 40 − (4 × 6) = 16.

(b) Under $X_1$ column:

Old element = 4
Entering column coefficient = 4
New pivot element = 1/2
Therefore, value = 4 − (4 × 1/2) = 2.

(c) Under $S_1$ column:

Old element = 1
Entering column coefficient = 4
New pivot element = 0
Therefore, value = $1 - (4 \times 0) = 1$.

(d) Under $S_2$ column:

Old element = 0
Entering column coefficient = 4
New pivot element = 1/10
Therefore, value = $0 - (4 \times 1/10) = -2/5$.

Now, enter these values in the $S_1$ row.

10. Calculate $Z_j = \sum C_B X_B = (0 \times 16) + (8 \times 6) = 48$.
11. Calculate $\Delta_j$ row values $= \sum C_B X_B - C_j$

$$\Delta_1 = [(0 \times 2) + (8 \times 1/2)] - 6 = -2$$
$$\Delta_2 = 0\,(\because \text{Pivotal column})$$
$$\Delta_3 = [(0 \times 1) + (8 \times 0)] - 0 = 0$$
$$\Delta_4 = [(0 \times -2/5) + (8 \times 1/10)] - 0 = 4/5$$

Now, identify the most negative value which is −2 under $X_1$ column. Mark that column with ↑ as shown, which is the entering variable.

12. Calculate the minimum ratio ($X_B/X_k$) for the BV rows only, i.e., $S_1$ but not $X_2$.

$$\text{Min ratio} = 16/2 = 8$$

Mark this row with (←), which marks $X_1$ as the entering variable that replaces $S_1$.

13. Identify the pivot element, which is the element at the interaction of $S_1$ row and $X_1$ columns, respectively. The pivot element is 2.
14. Proceed to the next iteration, where $X_1$ replaces $S_1$.

**Simplex Table 3**

| | | $C_j \rightarrow$ | 6 | 8 | 0 | 0 | |
|---|---|---|---|---|---|---|---|
| Basic Variable (B.V.) | $C_B$ | $X_B$ | $X_1$ | $X_2$ | $S_1$ | $S_2$ | Min ratio ($X_B/X_K$) |
| $X_1$ | 6 | 8 | 1 | 0 | 1/2 | - 1/5 | – |
| $X_2$ | 8 | 2 | 0 | 1 | - 1/4 | - 3/10 | – |
| | $Z_j = 64$ | | 0 | 0 | 1 | 6/5 | |
| | | | ↑ | | | | |

The values for $X_2$ row are calculated using formula 3.1 as follows:

(a) Under $X_B$ column:

Old element = 6
Entering column coefficient = 1/2
New pivot element = 8
Therefore, value = 6 − (1/2 × 8) = 2.

(b) Under $X_2$ column:

Old element = 1
Entering column coefficient = 1/2
New pivot element = 0
Therefore, value = 1 − (1/2 × 0) = 1.

(c) Under $S_1$ column:

Old element = 0
Entering column coefficient = 1/2
New pivot element = 0
Therefore, value = 0 − (1/2 × 1/2) = −1/4.

(d) Under $S_2$ column:

Old element = 1/10
Entering column coefficient = 1/2
New pivot element = −1/5
Therefore, value = 1/10 − (1/2 × −1/5) = 3/10.

15. The value of $Z_j = \sum C_B X_B = [(6 \times 8) + (8 \times 2)] = 64$.
16. The values for $\Delta_j$ row:

$$\Delta_1 = [(6 \times 1) + (8 \times 0)] - 6 = 0$$
$$\Delta_2 = [(6 \times 0) + (8 \times 1)] - 8 = 0$$
$$\Delta_3 = [(6 \times 1/2) + (8 \times -1/4)] - 0 = 1$$
$$\Delta_4 = [(6 \times -1/5) + (8 \times 3/10)] - 0 = 6/5$$

All the values in the $\Delta_{j+I}$ row are positive. Hence, optimality is reached, and there is no need for further iterations.
The solution should be represented as

$$Z = 64 \quad X_1 = 8 \quad X_2 = 2$$

where, the values for $X_1$ and $X_2$ should be taken from the $X_B$ column.
We suggest that the reader practice more problems using formula 3.1, which helps in getting a good hold on the formula, thus saving a lot of computation time in the exam.

## Tips to lessen computation time

1. Once the pivot element is identified, in the next immediate iteration, all elements except pivotal element are zeros in the pivotal column. The pivot element in the previous iteration will be replaced by 1 in the same place in the next immediate iteration.
2. If the new pivotal element is zero, rewrite the same values in the next iteration without computing anything.
   (In Simplex Table 2, 0 is the new pivot element for $S_1$. So, rewrite the value 1 under $S_1$ in Simplex Table 1 and in Simplex Table 2)
3. If the entering column coefficient is zero, all the values for that respective row remain unchanged in the next iteration also.

**Illustration 3.2**
Solve the following LP by simplex method

$$\begin{aligned} \text{Maximise } Z = {} & 3X_1 + 2X_2 + 5X_3 \\ \text{s.t.} \quad & X_1 + 2X_2 + X_3 \le 430 \\ & 3X_1 + 2X_3 \le 460 \\ & X_1 + 4X_2 \le 420 \\ & X_1, X_2, X_3 \ge 0 \end{aligned}$$

**Solution**
The standard form of the above LP will be

$$\begin{aligned} \text{Maximise } Z = {} & 3X_1 + 2X_2 + 5X_3 + 0.S_1 + 0.S_2 + 0.S_3 \\ \text{s.t.} \quad & X_1 + 2X_2 + X_3 + S_1 = 430 \\ & 3X_1 + 2X_3 + S_2 = 460 \\ & X_1 + 4X_2 + S_3 = 420 \\ & X_1, X_2, X_3, S_1, S_2, S_3 \ge 0 \end{aligned}$$

(*Note* There are no values for $X_2$ and $X_3$ in the second and third constraints, respectively. Therefore, the respective values will be 0 in the IBFS.)

**Simplex Table 1**

| | | $C_j$ → | 3 | 2 | 5 | 0 | 0 | 0 | |
|---|---|---|---|---|---|---|---|---|---|
| Basic Variable (B.V.) | $C_B$ | $X_B$ | $X_1$ | $X_2$ | $X_3$ | $S_1$ | $S_2$ | $S_3$ | Min ratio ($X_B/X_K$) |
| $S_1$ | 0 | 430 | 1 | 2 | 1 | 1 | 0 | 0 | 430/1 = 430 |
| $S_2$ | 0 | 460 | 3 | 0 | [2] | 0 | 1 | 0 | 460/2 = 230 ← |
| $S_3$ | 0 | 420 | 1 | 4 | 0 | 0 | 0 | 1 | – |
| | $Z_j = 0$ | | -3 | -2 | -5 | 0 | 0 | 0 | ← $\Delta_j$ |
| | | | | | ↑ Min $\Delta_j$ | | | | |

For the Min. ratio ($X_B/X_k$), it is mandatory that $X_k = 0$, but $X_B$ can be 0. Here, min $\Delta_j = -5$ under $X_3$ column, pivot element is 2, $X_3$ is the entering variable, and $S_2$ is the leaving variable.

Proceeding to the next iteration,

**Simplex Table 2**

| | | $C_j \rightarrow$ | 3 | 2 | 5 | 0 | 0 | 0 | |
|---|---|---|---|---|---|---|---|---|---|
| Basic Variable (B.V.) | $C_B$ | $X_B$ | $X_1$ | $X_2$ | $X_3$ | $S_1$ | $S_2$ | $S_3$ | Min ratio ($X_B/X_K$) |
| $S_1$ | 0 | 200 | -1/2 | 2 | 0 | 1 | -1/2 | 0 | 200/2 = 100← |
| $X_3$ | 5 | 230 | 3/2 | 0 | 1 | 0 | 1/2 | 0 | – |
| $S_3$ | 0 | 420 | 1 | 4 | 0 | 0 | 0 | 1 | 420/4 = 105 |
| | $Z_j = 1150$ | | 9/2 | - 2 ↑ | 0 | 0 | 5/2 | 0 | ← $\Delta_j$ |

Calculations for Simplex Table 2.

1. For $S_1$ row.

   (a) Under $X_B$: Value(V) $= 430 - (1 \times 230) = 200$
   (b) Under $X_1$: Value(V) $= 1 - (1 \times 3/2) = -1/2$
   (c) Under $X_2$: V $= 2 - (1 \times 0) = 2$
   (d) Under $S_1$: V $= 1 - (1 \times 0) = 1$
   (e) Under $S_2$: V $= 0 - (1 \times 1/2) = -1/2$
   (f) Under $S_3$: V $= 0 - (1 \times 0) = 0$

2. For $S_3$ row

   (a) Under $X_B$: V $= 420 - (0 \times 230) = 420$

   Note that the entering column coefficient for $S_3$ row is 0. Therefore, all the values in $S_3$ row in Simplex Table 1 can be directly transferred to Simplex Table 2.
   After calculations, from Simplex Table 2,

$$\text{Min } \Delta_j = -2, \text{under } X_2 \text{ column,}$$
$$\text{Min ratio} = 200/2 = 100$$

   Therefore, entering variable $= X_2$, leaving variable $= S_1$, and pivot element $= 2$, and $Z_j = 1{,}150$

Proceeding to the next iteration,

**Simplex Table 3**

| | | $C_j$ → | 3 | 2 | 5 | 0 | 0 | 0 | |
|---|---|---|---|---|---|---|---|---|---|
| Basic Variable (B.V.) | $C_B$ | $X_B$ | $X_1$ | $X_2$ | $X_3$ | $S_1$ | $S_2$ | $S_3$ | Min ratio ($X_B/X_K$) |
| $X_2$ | 2 | 100 | -1/4 | 1 | 0 | ½ | -1/4 | 0 | |
| $X_3$ | 5 | 230 | 3/2 | 0 | 1 | 0 | 1/2 | 0 | |
| $S_3$ | 0 | 20 | 2 | 0 | 0 | -2 | 1 | 1 | |
| | $Z_j = 1350$ | | 4 | 0 | 0 | 1 | 2 | 0 | ← $\Delta_j$ |
| | | | | ↑ | | | | | |

Calculations for Simplex Table 3

1. For $X_3$ row
   Since the entering column coefficient for $X_3$ row is 0 (below 2, the pivot element), all values for $X_3$ row remain the same.
2. For $S_3$ row

   (a) Under $X_B$: V = 420 − (4 × 100) = 20
   (b) Under $X_1$: V = 1 − (4 × −1/4) = 2
   (c) Under $X_3$: V = 0 − (4 × 0) = 0 ← new pivot element is zero. Hence value remains the same.
   (d) Under $S_1$: V = 0 − (4 × 1/2) = −2
   (e) Under $S_2$: V = 0 − (4 × −1/4) = 1
   (f) Under $S_3$: V = 1 − (4 × 0) = 1 ← New pivot element is zero.

**Illustration 3.3**
Solve the following LP by simplex method:

$$\begin{aligned}\text{Maximise } Z &= X_1 - 3X_2 + 2X_3\\ \text{s.t.} \quad & 3X_1 - X_2 + 3X_3 \leq 7\\ & -2X_1 + 4X_2 \leq 12\\ & -4X_1 + 3X_2 + 8X_3 \leq 10\\ & \text{and } X_1, X_2, X_3 \geq 0\end{aligned}$$

**Solution**
The objective function is of minimization type. The first step is to convert it to maximization type by multiplying the objective function by −1. Therefore, we get

$$\begin{aligned}\text{Maximize } Z' &= -X_1 + 3X_2 - 2X_3 + 0.S_1 + 0.S_2 + 0.S_3\\ \text{s.t.} \quad & 3X_1 - X_2 + 3X_3 + S_1 = 7\\ & -2X_1 + 4X_2 + S_2 = 12\\ & -4X_1 + 3X_2 + 8X_3 + S_3 = 10\\ & \text{and } X_1, X_2, X_3, S_1, S_2, S_3 \geq 0 \text{ and } Z' = -Z.\end{aligned}$$

The IBFS is as follows:

**Simplex Table 1**

| | | $C_j \rightarrow$ | -1 | 3 | -2 | 0 | 0 | 0 | |
|---|---|---|---|---|---|---|---|---|---|
| Basic Variable (B.V.) | $C_B$ | $X_B$ | $X_1$ | $X_2$ | $X_3$ | $S_1$ | $S_2$ | $S_3$ | Min ratio ($X_B/X_K$) |
| $S_1$ | 0 | 7 | 3 | -1 | 3 | 1 | 0 | 0 | – |
| $S_2$ | 0 | 12 | - 2 | $\boxed{4}$ | 0 | 0 | 1 | 0 | 12/4 = 3← |
| $S_3$ | 0 | 10 | - 4 | 3 | 8 | 0 | 0 | 1 | 10/3 = 3.33 |
| | $Z_j = 0$ | | + 1 | - 3 | + 2 | 0 | 0 | 0 | ← $\Delta_j$ |
| | | | | ↑ | | | | | |

**Simplex Table 2**

| | | $C_j \rightarrow$ | -1 | 3 | -2 | 0 | 0 | 0 | |
|---|---|---|---|---|---|---|---|---|---|
| Basic Variable (B.V.) | $C_B$ | $X_B$ | $X_1$ | $X_2$ | $X_3$ | $S_1$ | $S_2$ | $S_3$ | Min ratio ($X_B/X_K$) |
| $X_1$ | 0 | 10 | $\boxed{5/2}$ | 0 | 3 | 1 | 1/4 | 0 | 10/(5/2) = 4← |
| $X_2$ | 3 | 3 | - 1/2 | 1 | 0 | 0 | 1/4 | 0 | – |
| $S_3$ | 0 | 1 | - 5/2 | 0 | 8 | 0 | - 3/4 | 1 | – |
| | $Z_j = 9$ | | - 1/2 | 0 | 0 | 0 | 3/4 | 0 | ← $\Delta_j$ |
| | | | ↑ | | | | | | |

**Simplex Table 3**

| | | $C_j \rightarrow$ | -1 | 3 | -2 | 0 | 0 | 0 | |
|---|---|---|---|---|---|---|---|---|---|
| Basic Variable (B.V.) | $C_B$ | $X_B$ | $X_1$ | $X_2$ | $X_3$ | $S_1$ | $S_2$ | $S_3$ | Min ratio ($X_B/X_K$) |
| $X_1$ | - 1 | 4 | 1 | 0 | 6/5 | 2/5 | 1/10 | 0 | – |
| $X_2$ | 3 | 5 | 0 | 1 | 3/5 | 1/5 | 3/10 | 0 | – |
| $S_3$ | 0 | 11 | 0 | 0 | 11 | 1 | - 1/2 | 1 | – |
| | $Z_j = 11$ | | 0 | 0 | 19/5 | 3/5 | 9/10 | 0 | ← $\Delta_j$ |
| | | | ↑ | | | | | | |

Since all $\Delta_j \geq 0$, the optimal solution is obtained in Simplex Table 3.
The value of $Z' = 11$, $X_1 = 4$, $X_2 = 5$, $X_3 = 0$.
Therefore, the value of $Z = -11$

**Illustration 3.4**

Solve the following LP by simplex method:

$$\begin{aligned} \text{Maximize } Z &= -4X_1 + 2X_2 + X_3 \\ \text{s.t.} \quad & X_1 + X_2 + X_3 \leq 3 \\ & 2X_1 + X_2 + X_3 \leq 4 \\ & X_1 - X_2 \leq 0 \\ & X_1, X_2, X_3 \geq 0 \end{aligned}$$

**Solution**

Converting to maximization form

$$\begin{aligned} \text{Maximize } Z' &= 4X_1 - 2X_2 - 1X_3 + 0.S_1 + 0.S_2 + 0.S_3 \\ \text{s.t.} \quad & X_1 + X_2 + X_3 + S_1 = 3 \\ & 2X_1 + X_2 + X_3 + S_2 = 4 \\ & X_1 - X_2 + S_3 = 0 \\ & X_1, X_2, X_3, S_1, S_2, S_3 \geq 0 \end{aligned}$$

**Simplex Table 1**

| Basic Variable (B.V.) | $C_B$ | $X_B$ | $X_1$ | $X_2$ | $X_3$ | $S_1$ | $S_2$ | $S_3$ | Min ratio ($X_B/X_K$) |
|---|---|---|---|---|---|---|---|---|---|
| | | $C_j \rightarrow$ | 4 | -2 | -1 | 0 | 0 | 0 | |
| $S_1$ | 0 | 3 | 1 | 1 | 1 | 1 | 0 | 0 | 3/1 = 3 |
| $S_2$ | 0 | 4 | 2 | 2 | 1 | 0 | 1 | 0 | 4/2 = 2 ← |
| $S_3$ | 0 | 0 | [1] | -1 | 0 | 0 | 0 | 1 | 0/1 ← |
| | $Z_j = 0$ | | -4 ↑ | 2 | 1 | 0 | 0 | 0 | ← $\Delta_j$ |

**Simplex Table 2**

| Basic Variable (B.V.) | $C_B$ | $X_B$ | $X_1$ | $X_2$ | $X_3$ | $S_1$ | $S_2$ | $S_3$ | Min ratio ($X_B/X_K$) |
|---|---|---|---|---|---|---|---|---|---|
| | | $C_j \rightarrow$ | 4 | -2 | -1 | 0 | 0 | 0 | |
| $S_1$ | 0 | 3 | 0 | 2 | 1 | 1 | 0 | -1 | 3/2 = 1.5 |
| $S_2$ | 0 | 4 | 0 | [-4] | 1 | 0 | 1 | -2 | 4/4 = 1← |
| $X_1$ | 4 | 0 | 1 | -1 | 0 | 0 | 0 | 1 | – |
| | $Z_j = 0$ | | 0 | -2 ↑ | 1 | 0 | 0 | 4 | ← $\Delta_j$ |

**Simplex Table 3**

| Basic Variable (B.V.) | $C_B$ | $X_B$ | $X_1$ | $X_2$ | $X_3$ | $S_1$ | $S_2$ | $S_3$ | Min ratio ($X_B/X_K$) |
|---|---|---|---|---|---|---|---|---|---|
| | | $C_j \rightarrow$ | 4 | -2 | -1 | 0 | 0 | 0 | |
| $S_1$ | 0 | 1 | 0 | 0 | 1/2 | 1 | -1/2 | 0 | – |
| $X_2$ | -2 | 1 | 0 | 1 | 1/4 | 0 | 1/4 | -1/2 | – |
| $X_1$ | 4 | 1 | 1 | 0 | 1/4 | 0 | 1/4 | 1/2 | – |
| | $Z_j = 2$ | | 0 | 0 ↑ | 3/2 | 0 | 1/2 | 3 | ← $\Delta_j$ |

The value of $Z' = 2$, $X_1 = 1$, $X_2 = 1$, $X_3 = 0$.
Therefore, the value of $Z = -2$.

**Illustration 3.5**

A company expects a profit of Rs. 25 and Rs. 40, respectively, from its two products $X$ and $Y$. The material requirements for one unit are as shown below.

| Product | $A$ | $B$ | $C$ |
|---|---|---|---|
| $X$ | 1/2 | 4 | 1/5 |
| $Y$ | 1/4 | 7 | 1/5 |

Resources available are 2,200 units of material A, 28,000 units of material B, and 1,400 units of material C.

(a) Formulate the above as an LP and find the optimal solution.
(b) What are the levels of consumption of the materials?

**Solution**

(a) The formulation is as below:

$$\begin{aligned} \text{Maximize } Z = \ & 25X_1 + 40X_2 \\ \text{s.t.} \quad & \left(\tfrac{1}{2}X_1\right) + \left(\tfrac{1}{4}X_2\right) \leq 2,200 \\ & 4X_1 + 7X_2 \leq 28,000 \\ & \left(\tfrac{1}{5}X_1\right) + \left(\tfrac{1}{5}X_2\right) \leq 1,400 \\ & X_1, X_2 \geq 0 \end{aligned}$$

We introduce slack variables to constraints,

$$\begin{aligned} \text{Maximize } Z = \ & 25X_1 + 40X_2 + 0.S_1 + 0.S_2 + 0.S_3 \\ \text{s.t.} \quad & 1/2X_1 + 1/4X_2 + S_1 = 2,200 \\ & 4X_1 + 7X_2 + S_2 = 28,000 \\ & 1/5X_1 + 1/5X_2 + S_3 = 1,400 \\ & X_1, X_2, S_1, S_2, S_3 \geq 0 \end{aligned}$$

**Simplex Table 1**

| | | $C_j \rightarrow$ | 25 | 40 | 0 | 0 | 0 | |
|---|---|---|---|---|---|---|---|---|
| Basic Variable (B.V.) | $C_B$ | $X_B$ | $X_1$ | $X_2$ | $S_1$ | $S_2$ | $S_3$ | Min ratio ($X_B/X_K$) |
| $S_1$ | 0 | 2200 | 1/2 | 1/4 | 1 | 0 | 0 | 4400 |
| $S_2$ | 0 | 28000 | 4 | [7] | 0 | 1 | 0 | 4000← |
| $S_3$ | 0 | 1400 | 1/5 | 1/5 | 0 | 0 | 1 | 7000 |
| | $Z_j = 0$ | | -25 | -40 ↑ | 0 | 0 | 0 | ← $\Delta_j$ |

**Simplex Table 2**

| Basic Variable (B.V.) | $C_B$ | $X_B$ | $X_1$ | $X_2$ | $S_1$ | $S_2$ | $S_3$ | Min ratio $(X_B/X_K)$ |
|---|---|---|---|---|---|---|---|---|
| | | $C_j \rightarrow$ | 25 | 40 | 0 | 0 | 0 | |
| $S_1$ | 0 | 1200 | [5/14] | 0 | 1 | -1/28 | 0 | 1200/(5/14) = 3360← |
| $X_2$ | 40 | 4000 | 4/7 | 1 | 0 | 1/7 | 0 | – |
| $S_3$ | 0 | 600 | 3/35 | 0 | 0 | -1/35 | 1 | 600/(3/35) = 7000 |
| | $Z_j$ = 1600 | | -15/7 ↑ | 0 | 0 | 40/7 | 0 | ← $\Delta_j$ |

**Simplex Table 3**

| Basic Variable (B.V.) | $C_B$ | $X_B$ | $X_1$ | $X_2$ | $S_1$ | $S_2$ | $S_3$ | Min ratio $(X_B/X_K)$ |
|---|---|---|---|---|---|---|---|---|
| | | $C_j \rightarrow$ | 25 | 40 | 0 | 0 | 0 | |
| $X_1$ | 25 | 3360 | 1 | 0 | 14/5 | -1/10 | 0 | – |
| $X_2$ | 40 | 2080 | 0 | 1 | 18/5 | 1/5 | 0 | – |
| $S_3$ | 0 | 312 | 0 | 0 | 6/25 | -1/50 | 1 | – |
| | $Z_j$ = 167200 | | 0 | 40 ↑ | 134 | 11/2 | 0 | ← $\Delta_j$ |

The optimal solution is $Z =$ Rs. 1,67,200. $X_1 =$ 3,360, $X_2 =$ 2,080

(b) Levels of consumption

For Material A

$$\begin{aligned} \text{Used} &= (1/2)(3,360) + (1/4)(2,080) = 2,200 \text{ units} \\ \text{Total available} &= 2,200 \text{ units} \\ \text{Unused} &= 0 \text{ units} \end{aligned}$$

For Material B

$$\begin{aligned} \text{Used} &= 4(3,360) + 7(2,080) = 28,000 \text{ units} \\ \text{Total available} &= 28,000 \text{ units} \\ \text{Unused} &= 0 \text{ units} \end{aligned}$$

For Material C

$$\begin{aligned} \text{Used} &= (1/5)(3,360) + (1/5)(2,080) = 1,088 \text{ units} \\ \text{Total available} &= 1,400 \text{ units} \\ \text{Unused} &= 1,400 - 1,088 = 312 \text{ units} \end{aligned}$$

## 3.3 Big M Method

If any or some of the constraints are of the type "=" or "≥" then a new variable called "**artificial variable**" will have to be introduced in each of such constraints with a positive unit coefficient. If the objective function is a maximization function, then the coefficient of the artificial variable in the objective function should be −**M**, else it should be +**M** (**M** being a large value).

The artificial variables are only introduced to solve the problem. They should not reflect in the final solution. For this purpose only, we introduce −**M** (Maximization function) and +**M** (Minimization function).

**Illustration 3.6**
Solve by Big M method

$$\begin{aligned} \text{Maximize Z} = \; & X_1 + 2X_2 + 3X_3 - X_4 \\ \text{s.t.} \quad & X_1 + 2X_2 + 3X_3 = 15 \\ & 2X_1 + X_2 + 5X_3 = 20 \\ & X_1 + 2X_2 + X_3 + X_4 = 10 \\ & X_1, X_2, X_3, X_4 \geq 0 \end{aligned}$$

**Solution**
It is required to introduce artificial variables as constraint equations that have (=) signs.

(*Note* If constraint equations have (=) signs, only artificial variables will have to be introduced. If the signs are (≥), both artificial variables and surplus variables are to be introduced, with coefficients being +1 and −1, respectively).

The standard form of the above LP is

$$\begin{aligned} \text{Maximize Z} = \; & X_1 + 2X_2 + 3X_3 - X_4 - MA_1 - MA_2 \\ \text{s.t.} \quad & X_1 + 2X_2 + 3X_3 + A_1 = 15 \\ & 2X_1 + X_2 + 5X_3 + A_2 = 20 \\ & X_1 + 2X_2 + X_3 + X_4 = 10 \\ & X_1, X_2, X_3, X_4 \geq 0 \end{aligned}$$

In this problem, $X_4$ is treated as surplus variable that is already introduced in the question itself, related to the third constraint equation.

Proceeding further, we draw IBFS.

**Simplex Table 1**

| | | $C_j \rightarrow$ | 1 | 2 | 3 | -1 | -M | -M | |
|---|---|---|---|---|---|---|---|---|---|
| Basic Variable (B.V.) | $C_B$ | $X_B$ | $X_1$ | $X_2$ | $X_3$ | $X_4$ | $A_1$ | $A_2$ | Min ratio ($X_B/X_K$) |
| $A_1$ | -M | 15 | 1 | 2 | 3 | 0 | 1 | 0 | 15/3 = 5 |
| $A_2$ | -M | 20 | 2 | 1 | [5] | 0 | 0 | 1 | 20/5 = 4 ← |
| $X_4$ | -1 | 10 | 1 | 2 | 1 | 1 | 0 | 0 | 10/1 = 10 |
| | $Z_j$ = (-35$M$ -10) | | -3$M$ -2 | -3$M$-2 | -8$M$-4 | 0 | 0 | 0 | ← $\Delta_j$ |
| | | | | | ↑ *Min* $\Delta_j$ | | | | |

(*Note* If a variable already provides a starting solution, then the value for that element under the $C_B$ column will be the coefficient in the objective function. Here, $-1$ is the coefficient of $X_4$, and hence, value under $C_B$ column is $-1$. If surplus or slack variables are introduced to balance the constraints, the values remain zero.)

Proceeding further,

**Simplex Table 2**

| | $C_j$ → | | 1 | 2 | 3 | -1 | -M | -M | |
|---|---|---|---|---|---|---|---|---|---|
| Basic Variable (B.V.) | $C_B$ | $X_B$ | $X_1$ | $X_2$ | $X_3$ | $X_4$ | $A_1$ | $A_2$ | Min ratio ($X_B/X_K$) |
| $A_1$ | -M | 3 | -1/5 | 7/5 | 0 | 0 | 1 | -3/5 | 3/(7/5) = 15/7 ← |
| $X_3$ | 3 | 4 | 2/5 | 1/5 | 1 | 0 | 0 | +1/5 | 4 x 5 = 20 |
| $X_4$ | -1 | 6 | 3/5 | 9/5 | 0 | 1 | 0 | -1/5 | 6/(9/5) = 30/9 |
| | $Z_j = (-3M + 6)$ | | M/5 – 2/5 | $(-7/5)M$ + 16/5 | 0 | 0 | 0 | $(3/5)M$ + 3/5 | ← $\Delta_j$ |
| | | | | ↑ | | | | | |

**Simplex Table 3**

| | $C_j$ → | | 1 | 2 | 3 | -1 | -M | -M | |
|---|---|---|---|---|---|---|---|---|---|
| Basic Variable (B.V.) | $C_B$ | $X_B$ | $X_1$ | $X_2$ | $X_3$ | $X_4$ | $A_1$ | $A_2$ | Min ratio ($X_B/X_K$) |
| $X_2$ | 2 | 15/7 | -1/7 | 1 | 0 | 0 | 5/7 | -3/7 | – |
| $X_3$ | 3 | 25/7 | 3/7 | 0 | 1 | 0 | -1/7 | 2/7 | 25/3 |
| $X_4$ | -1 | 15/7 | 6/7 | 0 | 0 | 1 | -9/7 | 3/7 | 15/6 = 5/3← |
| | $Z_j = 90/7$ | | -30/35 | 0 | 0 | 0 | (16/7) + $M$ | $M$ – (3/7) | ← $\Delta_j$ |
| | | | ↑ | | | | | | |

**Simplex Table 4**

| | $C_j$ → | | 1 | 2 | 3 | 0 | -M | -M | |
|---|---|---|---|---|---|---|---|---|---|
| Basic Variable (B.V.) | $C_B$ | $X_B$ | $X_1$ | $X_2$ | $X_3$ | $X_4$ | $A_1$ | $A_2$ | Min ratio ($X_B/X_K$) |
| $X_2$ | 2 | 5/2 | 0 | 1 | 0 | 1/6 | 1/2 | -5/14 | – |
| $X_3$ | 3 | 5/2 | 0 | 0 | 1 | 3/6 | 1/2 | 1/14 | – |
| $X_1$ | 1 | 5/2 | 1 | 0 | 0 | 7/6 | -3/2 | 1/2 | – |
| | $Z_j = 15$ | | 0 | 0 | 0 | 75/36 | $M$ – 5/2 | $M$ + 15/28 | ← $\Delta_j$ |

The value of $Z = 15$, $X_1 = 5/2$, $X_2 = 5/2$, $X_3 = 5/2$, and all $\Delta_j \geq 0$.

**Illustration 3.7**

Use Charne's penalty method to solve the following LP:

$$\begin{aligned} \text{Maximize } Z = \ & 4X_1 + X_2 \\ \text{s.t.} \quad & 3X_1 + X_2 = 3 \\ & 4X_1 + 3X_2 \geq 6 \\ & X_1 + 2X_2 \leq 3 \\ & X_1, X_2 \geq 0 \end{aligned}$$

**Solution**

The problem is of minimization type. Converting it to maximization type, we get

$$\begin{aligned} \text{Maximize } Z' = \ & -4X_1 - X_2 + 0.S_1 + 0.S_2 - MA_1 - MA_2 \\ \text{s.t.} \quad & 3X_1 + X_2 + A_1 = 3 \\ & 4X_1 + 3X_2 - S_1 + A_2 = 6 \\ & X_1 + 2X_2 + S_2 = 3 \\ & X_1, X_2, S_1, S_2, A_1, A_2 \geq 0 \end{aligned}$$

**Simplex Table 1**

| | | $C_j \rightarrow$ | -4 | -1 | 0 | 0 | -M | -M | |
|---|---|---|---|---|---|---|---|---|---|
| Basic Variable (B.V.) | $C_B$ | $X_B$ | $X_1$ | $X_2$ | $S_1$ | $S_2$ | $A_1$ | $A_2$ | Min ratio ($X_B/X_K$) |
| $A_1$ | $-M$ | 3 | [3] | 1 | 0 | 0 | 1 | 0 | 3/3 = 1← |
| $A_2$ | $-M$ | 6 | 4 | 3 | -1 | 0 | 0 | 1 | 6/4 = 1.5 |
| $S_2$ | 0 | 3 | 1 | 2 | 0 | 1 | 0 | 0 | 3/1 = 3 |
| | $Z_j = -9M$ | | $-7M + 4$ ↑ | $-4M + 1$ | $M$ | 0 | 0 | 0 | ← $\Delta_j$ |

**Simplex Table 2**

| | | $C_j \rightarrow$ | -4 | -1 | 0 | 0 | -M | -M | |
|---|---|---|---|---|---|---|---|---|---|
| Basic Variable (B.V.) | $C_B$ | $X_B$ | $X_1$ | $X_2$ | $S_1$ | $S_2$ | $A_1$ | $A_2$ | Min ratio ($X_B/X_K$) |
| $X_1$ | -4 | 1 | 1 | 1/3 | 0 | 0 | 1/3 | 0 | – |
| $A_2$ | $-M$ | 2 | 0 | 5/3 | -1 | 0 | -4/3 | 1 | 6/5 |
| $S_2$ | 0 | 2 | 0 | [5/3] | 0 | 1 | -1/30 | 0 | 6/5← |
| | $Z_j = -4 - 2M$ | | 0 | $(-5M + 1)/3$ ↑ | $M$ | 0 | $(-4 + M)/3$ | 0 | ← $\Delta_j$ |

**Simplex Table 3**

| | | $C_j$ → | -4 | -1 | 0 | 0 | -M | -M | |
|---|---|---|---|---|---|---|---|---|---|
| Basic Variable (B.V.) | $C_B$ | $X_B$ | $X_1$ | $X_2$ | $S_1$ | $S_2$ | $A_1$ | $A_2$ | Min ratio $(X_B/X_K)$ |
| $X_1$ | -4 | 3/5 | 1 | 0 | 0 | -1/5 | 6/15 | 0 | – |
| $A_2$ | $-M$ | 0 | 0 | 0 | -1 | -1 | -1 | 1 | – |
| $X_2$ | -1 | 6/5 | 0 | 1 | 0 | 3/5 | -1/5 | 0 | – |
| | $Z_j$ = -18/5 | | 0 | 0 | $M$ | (1/5) + $M$ | $M$ – (27/15) | 0 | ← $\Delta_j$ |

Since all $\Delta_j \geq 0$, $Z' = -18/5$ or $Z = 18/5$, $X_1 = 3/5$, $X_2 = 6/5$.

## Illustration 3.8

Solve the following LP by Big M Technique:

$$\begin{aligned} \text{Maximize } Z = \; & 5X_1 - 2X_2 - X_3 \\ \text{s.t.} \quad & 2X_1 + 2X_2 - X_3 \geq 2 \\ & 3X_1 - 4X_2 \geq 3 \\ & X_2 + 3X_3 \geq 5 \\ & X_1, X_2, X_3 \geq 0 \end{aligned}$$

## Solution

The standard form will be

$$\begin{aligned} \text{Maximize } Z = \; & 5X_1 - 2X_2 - X_3 + 0.S_1 + 0.S_2 - MA_1 - MA_2 - MA_3 \\ \text{s.t.} \quad & 2X_1 + 2X_2 - X_3 + A_1 - S_1 \geq 2 \\ & 3X_1 - 4X_2 + A_2 - S_2 \geq 3 \\ & X_2 + 3X_3 + A_3 - S_3 \geq 5 \\ & X_1, X_2, X_3, S_1, S_2, S_3, A_1, A_2, A_3 \geq 0 \end{aligned}$$

**Simplex Table 1**

| | $C_j$ | → | 5 | -2 | -1 | 0 | 0 | 0 | $-M$ | $-M$ | $-M$ | |
|---|---|---|---|---|---|---|---|---|---|---|---|---|
| Basic Variable (B.V.) | $C_B$ | $X_B$ | $X_1$ | $X_2$ | $X_3$ | $S_1$ | $S_2$ | $S_3$ | $A_1$ | $A_2$ | $A_3$ | Min ratio $(X_B/X_K)$ |
| $A_1$ | $-M$ | 2 | 2 | 2 | -1 | -1 | 0 | 0 | 1 | 0 | 0 | 2/2=1 |
| $A_2$ | $-M$ | 3 | [3] | -4 | 0 | 0 | -1 | – | – | 1 | 0 | 3/3=1 ← |
| $A_3$ | $-M$ | 5 | 0 | 1 | 3 | 0 | 0 | -1 | 0 | 0 | 1 | – |
| | $Z_j$ = -10$M$ | | -5$M$ - 5 | 2$M$ + 2 | $M$ + 1 | $M$ | $M$ | $M$ | 0 | 0 | 0 | $\Delta_j$ ← |
| | | | ↑ | | | | | | | | | |

There is a tie for the pivotal row. This situation is termed as degeneracy in simplex methods. Refer Sect. 3.7.5, in order to comprehend the solution better.

**Simplex Table 2**

| Basic Variable (B.V.) | $C_B$ | $X_B$ | $X_1$ | $X_2$ | $X_3$ | $S_1$ | $S_2$ | $S_3$ | $A_1$ | $A_2$ | $A_3$ | Min ratio ($X_B/X_K$) |
|---|---|---|---|---|---|---|---|---|---|---|---|---|
| | $C_j$ → | | 5 | -2 | -1 | 0 | 0 | 0 | $-M$ | $-M$ | $-M$ | |
| $A_1$ | $-M$ | 0 | 0 | [14/3] | -1 | -1 | 2/3 | 0 | 1 | -2/3 | 0 | 0/(14/3) = 0 ← |
| $X_1$ | 5 | 1 | 1 | -4/3 | 0 | 0 | -1/3 | 0 | 0 | 1/3 | 0 | – |
| $A_3$ | $-M$ | 5 | 0 | 1 | 3 | 0 | 0 | -1 | 0 | 0 | 1 | 5/1=5 |
| | $Z_j = -5M + 5$ | | 0 | $(-14-17M)/3$ ↑ | $-2M+1$ | $M$ | $-(2M/3)-(5/3)$ | $M$ | 0 | $5M/3$ | 0 | $\Delta_j$ ← |

**Simplex Table 3**

| Basic Variable (B.V.) | $C_B$ | $X_B$ | $X_1$ | $X_2$ | $X_3$ | $S_1$ | $S_2$ | $S_3$ | $A_1$ | $A_2$ | $A_3$ | Min ratio ($X_B/X_K$) |
|---|---|---|---|---|---|---|---|---|---|---|---|---|
| | $C_j$ → | | 5 | -2 | -1 | 0 | 0 | 0 | $-M$ | $-M$ | $-M$ | |
| $X_2$ | -2 | 0 | 0 | 1 | -3/14 | -3/14 | 1/7 | 0 | 3/14 | -1/7 | 0 | – |
| $X_1$ | 5 | 1 | 1 | 0 | -2/7 | -2/7 | -1/7 | 0 | 2/7 | 1/7 | 0 | – |
| $A_3$ | $-M$ | 5 | 0 | 0 | [45/14] | 3/14 | -1/7 | -1 | -3/14 | 1/7 | 1 | 5/(45/14) ← |
| | $Z_j = -5M + 5$ | | 0 | 0 | $-45M/14$ ↑ | $-1-(3/14)M$ | $-1+(M/7)$ | $M$ | $1+(17M/14)$ | $1-(6M/7)$ | 0 | $\Delta_j$ ← |

**Simplex Table 4**

| Basic Variable (B.V.) | $C_B$ | $X_B$ | $X_1$ | $X_2$ | $X_3$ | $S_1$ | $S_2$ | $S_3$ | $A_1$ | $A_2$ | $A_3$ | Min ratio ($X_B/X_K$) |
|---|---|---|---|---|---|---|---|---|---|---|---|---|
| | $C_j$ → | | 5 | -2 | -1 | 0 | 0 | 0 | $-M$ | $-M$ | $-M$ | |
| $X_2$ | -2 | 1/3 | 0 | 1 | 0 | -1/5 | 2/15 | -1/15 | 1/5 | -2/65 | 7/15 | – |
| $X_1$ | 5 | 13/9 | 1 | 0 | 0 | -4/15 | -7/45 | -4/45 | 4/15 | 7/45 | 4/45 | – |
| $X_3$ | -1 | 14/9 | 0 | 0 | 1 | 1/15 | -2/45 | -14/45 | -1/15 | 2/45 | 14/45 | – |
| | $Z_j = 5$ | | 0 | 0 | 0 | 1 | 1 | 0 | $M+1$ | $M+7/15$ | $M$ | $\Delta_j$ ← |

Since all $\Delta_j \geq 0$, the optimal solution will be $Z = 5$, $X_1 = 13/9$, $X_2 = 1/3$, $X_3 = 14/9$

## *3.3.1 Unrestricted Variables*

This solution arrives when no constraints are**. Such a difficulty can be removed using the transformation

$$X_1 = X_1' - X_1'', \quad \text{and} \quad X_2 = X_2' - X_2'', \quad \text{where } X_1', X_1'', X_2', X_2'' \geq 0$$

We will consider a problem to explain this further.

**Illustration 3.9**
Solve the following LP

$$\begin{aligned} \text{Maximize } Z = \; & 2X_1 + 3X_2 \\ \text{s.t.} \quad & -X_1 + 3X_2 \leq 4 \\ & X_1 + X_2 \leq 6 \\ & X_1 + 3X_2 \leq 9 \\ & \text{and } X_1, X_2 \text{ unrestricted} \end{aligned}$$

**Solution**
We replace $X_1$ and, $X_2$ as

$$X_1 = X_1' - X_1'' \quad \text{and} \quad X_2 = X_2' - X_2''$$

Now, the problem can be written as

$$\begin{aligned} \text{Maximize } Z = \; & 2(X_1' - X_1'') + 3(X_1' - X_1'') \\ \text{s.t.} \quad & -X_1' + X_1'' + 2X_2' - 2X_2'' \leq 4 \\ & X_1' - X_1'' + X_2' - X_2'' \leq 6 \\ & X_1' - X_1'' + 3X_2' - 3X_2'' \leq 9 \\ & \text{and} \quad X_1', X_1'', X_2', X_2'' \geq 0 \end{aligned}$$

The standard form can now be written as

$$\begin{aligned} \text{Maximize } Z = \; & 2X_1' - 2X_1'' + 3X_1' - 3X_1'' + 0.S_1 + 0.S_2 + 0.S_3 \\ \text{s.t.} \quad & -X_1' + X_1'' + 2X_2' - 2X_2'' + S_1 = 4 \\ & X_1' - X_1'' + X_2' - X_2'' + S_2 = 6 \\ & X_1' - X_1'' + 3X_2' - 3X_2'' + S_3 = 9 \\ & \text{and } X_1', X_1'', X_2', X_2'', S_1, S_2, S_3 \geq 0 \end{aligned}$$

Now, solve by usual simplex method.

**Simplex Table 1**

| | | $C_j$ → | 2 | -2 | 3 | -3 | 0 | 0 | 0 | |
|---|---|---|---|---|---|---|---|---|---|---|
| Basic Variable (B.V.) | $C_B$ | $X_B$ | $X'_1$ | $X''_1$ | $X'_2$ | $X''_2$ | $S_1$ | $S_2$ | $S_3$ | Min ratio ($X_B/X_K$) |
| $S_1$ | 0 | 4 | -1 | 1 | [2] | -2 | 1 | 0 | 0 | 4/2=2← |
| $S_2$ | 0 | 6 | 1 | -1 | 1 | -1 | 0 | 1 | 0 | 6/1=6 |
| $S_3$ | 0 | 9 | 1 | -1 | 3 | -3 | 0 | 0 | 1 | 9/3=3 |
| | $Z_j = 0$ | | -2 | 2 | -3 ↑ | 3 | 0 | 0 | 0 | $\Delta_j$ ← |

**Simplex Table 2**

| | $C_j \rightarrow$ | | 2 | -2 | 3 | -3 | 0 | 0 | 0 | |
|---|---|---|---|---|---|---|---|---|---|---|
| Basic Variable (B.V.) | $C_B$ | $X_B$ | $X'_1$ | $X''_1$ | $X'_2$ | $X''_2$ | $S_1$ | $S_2$ | $S_3$ | Min ratio ($X_B/X_K$) |
| $X'_2$ | 3 | 2 | -1/2 | 1/2 | 1 | -1 | 1/2 | 0 | 0 | – |
| $S_2$ | 0 | 4 | 3/2 | -3/2 | 0 | 0 | -1/2 | 1 | 0 | 8/3 |
| $S_3$ | 0 | 3 | [5/2] | -5/2 | 0 | 0 | -3/2 | 0 | 1 | 6/5← |
| | $Z_j = 6$ | | -7/2 | 7/2 | 0 | 0 | 3/2 | 0 | 0 | $\Delta_j$ ← |
| | | | ↑ | | | | | | | |

**Simplex Table 3**

| | $C_j \rightarrow$ | | 2 | -2 | 3 | -3 | 0 | 0 | 0 | |
|---|---|---|---|---|---|---|---|---|---|---|
| Basic Variable (B.V.) | $C_B$ | $X_B$ | $X'_1$ | $X''_1$ | $X'_2$ | $X''_2$ | $S_1$ | $S_2$ | $S_3$ | Min ratio ($X_B/X_K$) |
| $X'_2$ | 3 | 13/5 | 0 | 0 | 1 | -1 | 1/5 | 0 | 1/5 | – |
| $S_2$ | 0 | 11/5 | 0 | 0 | 0 | 0 | [2/5] | 1 | -3/5 | 11/2 |
| $X'_1$ | 2 | 6/5 | 1 | -1 | 0 | 0 | -3/5 | 0 | 2/5 | – |
| | $Z_j = 51/5$ | | 0 | 0 | 0 | 0 | -3/5 | 0 | 0 | $\Delta_j$ ← |
| | | | | | | | ↑ | | | |

**Simplex Table 4**

| | $C_j \rightarrow$ | | 2 | -2 | 3 | -3 | 0 | 0 | 0 | |
|---|---|---|---|---|---|---|---|---|---|---|
| Basic Variable (B.V.) | $C_B$ | $X_B$ | $X'_1$ | $X''_1$ | $X'_2$ | $X''_2$ | $S_1$ | $S_2$ | $S_3$ | Min ratio ($X_B/X_K$) |
| $X'_2$ | 3 | 3/2 | 0 | 0 | 1 | -1 | 0 | 0 | 1/5 | – |
| $S_2$ | 0 | 11/2 | 0 | 0 | 0 | 0 | 1 | 5/2 | -2/3 | – |
| $X'_1$ | 2 | 9/2 | [1] | -1 | 0 | 0 | 0 | 0 | 2/5 | – |
| | $Z_j = 27/2$ | | 0 | 2 | 0 | 0 | 0 | 0 | 0 | $\Delta_j$ ← |
| | | | ↑ | | | | | | | |

Optimality has been obtained as all $\Delta_j \geq 0$.

$$Z_j = 27/2 \quad X_1' = 9/2 \quad X_2' = 3/2 \quad X_1'' = 0 \quad X_2'' = 0$$

$$X_1 = X_1' - X_1'' = 9/2 - 0 = 9/2 \quad X_2 = X_2' - X_2'' = 3/2 - 0 = 3/2$$

Therefore, the optimal values are $Z = 27/2$, $X_1 = 9/2$, $X_2 = 3/2$.

**Illustration 3.10**

$$\begin{aligned} \text{Maximize } Z = \;& 8X_2 \\ \text{s.t.} \quad & X_1 - X_2 \geq 0 \\ & 2X_1 + 3X_2 \leq -6 \\ & \text{and } X_1, X_2 \text{ unrestricted} \end{aligned}$$

**Solution**

The standard form is as follows:

$$\begin{aligned} \text{Maximize } Z = \;& 0X_1' + 0X_1'' + 8\left(X_2' - X_2''\right) - MA_1 - MA_2 + 0.S_1 + 0.S_2 \\ \text{s.t.} \quad & X_1' - X_1'' - X_2' + X_2'' - S_1 + A_1 = 0 \\ & 2X_1' + 2X_1'' - 3X_2' + 3X_2'' - S_2 + A_2 = 6 \\ & X_1', X_1'', X_2', X_2'', S_1, S_2, A_1, A_2 = 0 \end{aligned}$$

The right-hand side of the second constraint has a negative sign. This has to be made positive. So, we multiply the equation by $(-1)$, where the $(\leq)$ sign becomes $(\geq)$. Now, artificial and surplus variables are added.

Proceeding further,

**Simplex Table 1**

| | | $C_j$ → | 0 | 0 | 8 | -8 | -M | -M | 0 | 0 | |
|---|---|---|---|---|---|---|---|---|---|---|---|
| Basic Variable (B.V.) | $C_B$ | $X_B$ | $X'_1$ | $X''_1$ | $X'_2$ | $X''_2$ | $A_1$ | $A_2$ | $S_1$ | $S_2$ | Min ratio ($X_B/X_K$) |
| $A_1$ | -M | 0 | 1 | 1 | 1 | [1] | 1 | 0 | -1 | 0 | 0/1← |
| $A_2$ | -M | 6 | -2 | 2 | -3 | 3 | 0 | 1 | 0 | -1 | 6/3 = 2 |
| | $Z_j$ = -6M | | M | -M | 4M -8 | -4M+8 ↑ | 0 | 1 | M | M | $\Delta_j$ ← |

**Simplex Table 2**

| | | $C_j$ → | 0 | 0 | 8 | -8 | -M | -M | 0 | 0 | |
|---|---|---|---|---|---|---|---|---|---|---|---|
| Basic Variable (B.V.) | $C_B$ | $X_B$ | $X'_1$ | $X''_1$ | $X'_2$ | $X''_2$ | $A_1$ | $A_2$ | $S_1$ | $S_2$ | Min ratio ($X_B/X_K$) |
| $X''_2$ | -8 | 0 | 1 | -1 | -1 | 1 | 1 | 0 | -1 | 0 | – |
| $A_2$ | -M | 6 | -5 | [5] | 0 | 0 | -3 | 1 | 3 | -1 | 6/5← |
| | $Z_j$ = -6M | | 5M-8 | -5M+8 ↑ | 0 | 0 | 4M – 8 | 0 | -3M+8 | M | $\Delta_j$ ← |

**Simplex Table 3**

| | | $C_j \rightarrow$ | 0 | 0 | 8 | -8 | $-M$ | $-M$ | 0 | 0 | |
|---|---|---|---|---|---|---|---|---|---|---|---|
| Basic Variable (B.V.) | $C_B$ | $X_B$ | $X'_1$ | $X''_1$ | $X'_2$ | $X''_2$ | $A_1$ | $A_2$ | $S_1$ | $S_2$ | Min ratio ($X_B/X_K$) |
| $X''_2$ | -8 | 6/5 | 0 | 0 | -1 | 1 | 2/5 | 1/5 | -2/5 | 1/5 | – |
| $X''_1$ | 0 | 6/5 | -1 | 1 | 0 | 0 | -3/5 | 1/5 | 3/5 | -1/5 | – |
| | $Z_j = -48/5$ | | 0 | 0 | 0 | 0 | $M-(16/5)$ | $M-(8/5)$ | 16/5 | 8/5 | $\Delta_j \leftarrow$ |

We find all $\Delta_j \geq 0$ from $\Delta_j$ row with

$$X''_1 = 6/5 \quad X''_2 = 6/5 \quad X'_1 = 0 \quad X'_2 = 0$$

Therefore,

$$X_1 = X'_1 - X''_1 = 0 - 6/5 = -6/5$$
$$X_2 = X'_1 - X''_2 = 0 - 6/5 = -6/5$$

Therefore, $Z = -48/5 \quad X_1 = 6/5 \quad X_2 = -6/5$

### 3.3.2 Introduction of a Constant

Suppose a constant is included in the objective function, then it should be deleted in the beginning and adjusted in the finally in the optimal value of $Z$. If there is an equality in one of the constraints, one variable can be eliminated from the inequalities with "$\leq$" or "$\geq$." All these are explained by a problem.

**Illustration 3.11**

$$\begin{aligned} \text{Maximize } Z = \; & 4X_1 + 5X_2 - 3X_3 + 50 \\ \text{s.t.} \quad & X_1 + X_2 + X_3 = 10 \\ & X_1 - X_2 \geq 1 \\ & X_1 + 3X_2 + X_3 \leq 40 \\ & X_1, X_2, X_3 \geq 0 \end{aligned}$$

Initially, we eliminate $X_3$ from latest constraint equation by subtracting from first constraint equation, and treat $X_3$ as slack variable in the first constraint equation. The LP can be represented in its standard form as below:

$$\begin{aligned} \text{Maximize } Z = \; & 4X_1 + 5X_2 - 3X_3 + 0.S_1 + 0.S_2 - MA_1 \\ \text{s.t.} \quad & X_1 + X_2 + X_3 = 10 \\ & X_1 - X_2 - S_1 + A_1 = 1 \\ & X_1 + 2X_2 + S_2 = 30 \end{aligned}$$

Now, we solve it by normal simplex method

**Simplex Table 1**

| | | $C_j \rightarrow$ | 4 | 5 | -3 | 0 | 0 | $-M$ | |
|---|---|---|---|---|---|---|---|---|---|
| Basic Variable (B.V.) | $C_B$ | $X_B$ | $X_1$ | $X_2$ | $X_3$ | $S_1$ | $S_2$ | $A_1$ | Min ratio $(X_B/X_K)$ |
| $X_3$ | -3 | 10 | 1 | 1 | 1 | 0 | 0 | 0 | 10/1=10 |
| $A_1$ | $-M$ | 1 | 1 | 1 | 0 | -1 | 0 | 1 | 1/1=1← |
| $S_2$ | 0 | 30 | 1 | 2 | 0 | 0 | 1 | 0 | 30/1=30 |
| | $Z_j = -M - 30$ | | $-7 - M$ | $-8 + M$ | 0 | $M$ | 0 | 0 | $\Delta_j$ ← |
| | | | ↑ | | | | | | |

**Simplex Table 2**

| | | $C_j \rightarrow$ | 4 | 5 | -3 | 0 | 0 | $-M$ | |
|---|---|---|---|---|---|---|---|---|---|
| Basic Variable (B.V.) | $C_B$ | $X_B$ | $X_1$ | $X_2$ | $X_3$ | $S_1$ | $S_2$ | $A_1$ | Min ratio $(X_B/X_K)$ |
| $X_3$ | -3 | 9 | 0 | 2 | 1 | 1 | 0 | 0 | 9/2← |
| $X_1$ | 4 | 1 | 1 | -1 | 0 | -1 | 0 | 0 | – |
| $S_2$ | 0 | 29 | 0 | 3 | 0 | 1 | 1 | 1 | 29/3 |
| | $Z_j = 9/2$ | | 0 | -15 | 0 | -7 | 0 | $M$ | $\Delta_j$ ← |
| | | | | ↑ | | | | | |

**Simplex Table 3**

| | | $C_j \rightarrow$ | 4 | 5 | -3 | 0 | 0 | $-M$ | |
|---|---|---|---|---|---|---|---|---|---|
| Basic Variable (B.V.) | $C_B$ | $X_B$ | $X_1$ | $X_2$ | $X_3$ | $S_1$ | $S_2$ | $A_1$ | Min ratio $(X_B/X_K)$ |
| $X_2$ | 5 | 9/2 | 0 | 1 | 1/2 | 1/2 | 0 | 0 | – |
| $X_1$ | 4 | 11/2 | 1 | 0 | 1/2 | -1/2 | 0 | 0 | – |
| $S_2$ | 0 | 31/2 | 0 | 0 | -3/2 | -1/2 | 1 | 1 | – |
| | $Z_j = 89/2$ | | 0 | 0 | 15/2 | 1/2 | 0 | $M$ | $\Delta_j$ ← |
| | | | | ↑ | | | | | |

We notice all $\Delta_j \geq 0$. So, the optimum values are $X_1 = 11/2$, $X_2 = 9/2$, $Z = 89/2 + 50 = 189/2$.

### *3.3.3 Disadvantages of Big M Method*

- Since the value of M is large, it makes computability inconvenient. This can be eliminated by a two-phase simplex method.
- Another difficulty is M should be assigned to values $C_1$, $C_2$ … in the objective function. This is difficult since a computer has a fixed number of digits.

## 3.4 Duality

If in any LPP the constraints are too many, then the time required to sole it increases. In such cases, by treating the original LPP as "primal," it can be converted to its "dual" LPP which takes lesser time to solve. Duality states that there is a related problem of maximization or minimization for every LP problem of minimization or maximization based on the same data and numerical values of the objective function of the two problems. The original problem is called the "primal" problem, and the other is called the "dual" problem. The simplex table directly provides all the primal-dual calculations. Hence, it is necessary to define the dual consistent with the standard form of the primal. We can obtain the dual according to the following rules:

There is a dual problem for every primal constraint.
There is a dual constraint for every primal variable.

The constraint coefficients of a primal variable form left-side coefficients of the corresponding dual constraints, and the objective functions' coefficients of the primal variable become the right-hand side of the dual constraint.

| Standard Primal Objective | Dual | | |
|---|---|---|---|
| | Objective | Constraint | Variable |
| Maximization | Maximization | $\geq$ | Unrestricted |
| Minimization | Minimization | $\leq$ | Unrestricted |

This can be illustrated by the following examples.

(*Note* An optimal solution for the "dual" is also an optimal solution for the "primal.")

**Illustration 3.12**
Construct the dual of the following LP:

$$\begin{aligned} \text{Maximize } Z = {} & 5X_1 + 3X_2 \\ \text{s.t.} \quad & 6X_1 + 10X_2 \leq 30 \\ & 10X_1 + 4X_2 \leq 20 \\ & X_1, X_2 \geq 0 \end{aligned}$$

**Solution**
The first step is to convert the given problem into its standard form, which is called standard primal. Therefore, the process becomes

$$\begin{aligned} \text{Maximize } Z_X = {} & 5X_1 + 3X_2 + 0.S_1 + 0.S_2 \\ \text{s.t.} \quad & 6X_1 + 10X_2 + S_1 = 30 \\ & 10X_1 + 4X_2 + S_2 = 20 \\ & X_1, X_2, S_1, S_2 \geq 0 \end{aligned}$$

Now, the dual can be written as follows:

$$\begin{aligned} \text{Maximize } Z_W = {} & 30w_1 + 20w_2 \\ \text{s.t.} \quad & 6w_1 + 10w_2 \geq 5 \\ & 10w_1 + 4w_2 \geq 3 \\ & 1w_1 + 0w_2 \geq 5 \\ & 0w_1 + 1w_2 \geq 5 \\ & \text{and } w_1, w_2 \geq 0 \end{aligned}$$

The right-hand constants in the primal problem have to be summed, and this forms the objective function for the dual problem.

For writing the constraints in the dual, sum all $X_1$ coefficients in the two constraint equations, and similarly, the $X_2$ coefficients $w_1$ and $w_2$ are written in place of $X_1$ and $X_2$. It is also necessary to sum up the slack ($S_1$ and $S_2$) components in both constraints of the standard primal.

The constraint inequality in the dual problem is always ($\geq$), and the variables ($w_1$ and $w_2$) will have to be mentioned as $\geq 0$.

**Illustration 3.13**

Write the dual for the following LP:

$$\begin{aligned} \text{Maximize } Z = {} & 2X_1 + 5X_2 + 15X_3 \\ \text{s.t.} \quad & X_1 + 2X_2 + 4X_3 \leq 15 \\ & 2X_1 + 5X_2 + 4X_3 \leq 20 \\ & 3X_1 + 4X_2 + X_3 \leq 25 \\ & X_1, X_2, X_3 \geq 20 \end{aligned}$$

**Solution**

The standard primal is as below,

$$\begin{aligned} \text{Maximize } Z = {} & 2X_1 + 5X_2 + 15X_3 + 0.S_1 + 0.S_2 + 0.S_3 \\ \text{s.t.} \quad & X_1 + 2X_2 + 4X_3 + S_1 \leq 15 \\ & 2X_1 + 5X_2 + 4X_3 + S_2 \leq 20 \\ & 3X_1 + 4X_2 + X_3 + S_3 \leq 25 \\ & X_1, X_2, X_3, S_1, S_2, S_3 \geq 20 \end{aligned}$$

The dual problem would be

$$\begin{aligned} \text{Maximize } Z_w = {} & 15w_1 + 20w_2 + 25w_3 \\ \text{s.t.} \quad & w_1 + 2w_2 + 3w_3 \geq 2 \\ & 2w_1 + 5w_2 + 4w_3 \geq 5 \\ & 4w_1 + 4w_2 + w_3 \geq 15 \\ & 1w_1 + 0w_2 + 0w_3 \geq 0 \\ & 0w_1 + 1w_2 + 0w_3 \geq 0 \\ & 0w_1 + 0w_2 + 1w_3 \geq 0 \\ & \text{and } w_1, w_2, w_3 \text{ unrestricted} \end{aligned}$$

**Illustration 3.14**
Write the dual of the following LP:

$$\begin{aligned} \text{Maximize } Z_x = {} & 4X_2 + 10X_3 \\ \text{s.t.} \quad & 2X_1 + 4X_2 \geq 4 \\ & 2X_1 + 2X_1 + 4X_3 \leq 6 \\ & 4X_1 - 2X_2 + 6X_3 = 8 \\ & X_1, X_2, X_3 \geq 0 \end{aligned}$$

**Solution**
First, convert the objective function to maximization type. Whenever a constraint has (=) sign straight away, two constraint equations with (≥) and (≤) have to be formed out of that respective equation, in order to form the dual. All (≥) signs will have to be made (≤) before the dual by multiplying by −1.

$$\begin{aligned} \text{Maximize } Z'_x = {} & -4X_2 - 10X_3 \\ \text{s.t.} \quad & -2X_1 - 4X_2 \leq -4 \\ & 2X_1 + 2X_2 + 4X_3 \leq 6 \\ & 4X_1 - 2X_2 + 6X_3 \leq 8 \\ & 4X_1 - 2X_2 + 6X_3 \geq 8 \text{or} -4X_1 + 2X_2 - 6X_3 \leq -8 \end{aligned}$$

The dual of the LP is

$$\begin{aligned} \text{Maximize } Z'_W = {} & 4w_1 + 6w_2 + 8w'_3 - 8w''_3 \\ \text{s.t.} \quad & -2w_1 + 2w_2 + 4w'_3 - 4w''_3 \geq 0 \\ & -4w_1 + 2w_2 - 2w'_3 + 2w''_3 \geq -4 \\ & 0w_1 + 4w_2 + 6w'_3 - 6w''_3 \geq -10 \\ & w_1, w_2, w_3 \geq 0 \end{aligned}$$

## 3.5 Application of Duality

We explain the concept of application of duality by taking two examples.

**Illustration 3.15**
Solve the following LP and write the solution for its dual problem through the final simplex table:

$$\begin{aligned} \text{Maximize } Z = {} & 40X_1 + 25X_2 + 50X_3 \\ \text{s.t.} \quad & X_1 + 2X_2 + X_3 \leq 36 \\ & 2X_1 + X_2 + 4X_3 \leq 60 \\ & 2X_1 + 5X_2 + X_3 \leq 45 \\ & X_1, X_2, X_3 \geq 0 \end{aligned}$$

**Solution**

Initial approach is to solve by usual simplex method.

$$\begin{aligned} \text{Maximize } Z &= 40X_1 + 25X_2 + 50X_3 + 0.S_1 + 0.S_2 + 0.S_3 \\ \text{s.t.} \quad & X_1 + 2X_2 + X_3 + S_1 = 36 \\ & 2X_1 + X_2 + 4X_3 + S_2 = 60 \\ & 2X_1 + 5X_2 + X_3 + S_3 = 45 \\ & X_1, X_2, X_3, S_1, S_2, S_3 \geq 0 \end{aligned}$$

**Simplex Table 1**

| | $C_j \rightarrow$ | | 40 | 25 | 50 | 0 | 0 | 0 | |
|---|---|---|---|---|---|---|---|---|---|
| Basic Variable (B.V.) | $C_B$ | $X_B$ | $X_1$ | $X_2$ | $X_3$ | $S_1$ | $S_2$ | $S_3$ | Min ratio $(X_B/X_K)$ |
| $S_1$ | 0 | 36 | 1 | 2 | 1 | 1 | 0 | 0 | 36/1=36 |
| $S_2$ | 0 | 60 | 2 | 1 | [4] | 0 | 1 | 0 | 60/4=15← |
| $S_3$ | 0 | 45 | 2 | 5 | 1 | 0 | 0 | 1 | 45/1=45 |
| | $Z_j = 0$ | | -40 | -25 | -50 | 0 | 0 | 0 | $\Delta_j$ ← |
| | | | | | ↑ | | | | |

**Simplex Table 2**

| | $C_j \rightarrow$ | | 40 | 25 | 50 | 0 | 0 | 0 | |
|---|---|---|---|---|---|---|---|---|---|
| Basic Variable (B.V.) | $C_B$ | $X_B$ | $X_1$ | $X_2$ | $X_3$ | $S_1$ | $S_2$ | $S_3$ | Min ratio $(X_B/X_K)$ |
| $S_1$ | 0 | 21 | 1/2 | 7/4 | 0 | 1 | -1/4 | 0 | 21/(1/2) = 42 |
| $X_2$ | 50 | 15 | 1/2 | 1/4 | 1 | 0 | 1/4 | 0 | – |
| $S_3$ | 0 | 30 | [3/2] | 19/4 | 0 | 0 | -1/4 | 1 | 30/(3/2)=20← |
| | $Z_j = 750$ | | -15 | -25/2 | 0 | 0 | 25/2 | 0 | $\Delta_j$ ← |
| | | | | ↑ | | | | | |

**Simplex Table 3**

| | $C_j \rightarrow$ | | 40 | 25 | 50 | 0 | 0 | 0 | |
|---|---|---|---|---|---|---|---|---|---|
| Basic Variable (B.V.) | $C_B$ | $X_B$ | $X_1$ | $X_2$ | $X_3$ | $S_1$ | $S_2$ | $S_3$ | Min ratio $(X_B/X_K)$ |
| $S_1$ | 0 | 11 | 0 | 1/6 | 0 | 1 | -1/6 | -1/3 | – |
| $X_3$ | 50 | 5 | 0 | -4/3 | 1 | 0 | 1/3 | -1/3 | – |
| $X_1$ | 40 | 20 | 1 | 19/6 | 0 | 0 | -1/6 | 2/3 | – |
| | $Z_j = 1050$ | | 0 | 35 | 0 | 0 | 10 | 10 | $\Delta_j$ ← |

As all $\Delta_j \geq 0$, the optimal solution will be $Z = 1{,}050, X_1 = 20, X_2 = 0, X_3 = 5$
Now, the dual problem can be written as

$$\begin{aligned} \text{Maximize } Z_W &= 36w_1 + 60w_2 + 45w_3 \\ \text{s.t.} \quad & w_1 + 2w_2 + 2w_3 \geq 40 \\ & 2w_1 + w_2 + 5w_3 \geq 25 \\ & w_1 + 4w_2 + w_3 \geq 0 \\ & w_1, w_2, w_3 \geq 0 \end{aligned}$$

The solution for the dual problem is obtained from the values under the slack variable columns on the $\Delta_j$ row of the final simplex table. We write the solution of the dual problem as

$$w_1 = 0 \quad w_2 = 10 \quad w_3 = 10 \quad Z_W = 1,050.$$

## 3.6 Dual Simplex Method (DSM)

If DSM is a specialized form of Simplex Method in which optimality is maintained in all the iterations, initially, it may not be a feasible solution, but successive iterations overcome this. When the solution becomes (in an iteration), the procedure will be stopped since the solution obtained is both feasible and optimal. This is also a subroutine for integer programming, where it is used to remove infeasibility due to additional constraints known as Gomory's constraints.

### *3.6.1 Dual Simplex Method Is Explained by an Example*

**Illustration 3.16**

Formulate the dual for the following LP and hence find the values of the primal problem by solving the dual:

$$\begin{aligned} \text{Maximize } Z &= 40X_1 + 50X_2 \\ \text{s.t.} \quad & 2X_1 + 3X_2 \leq 3 \\ & 8X_1 + 4X_2 \leq 5 \\ & X_1, X_2 \geq 0 \end{aligned}$$

**Solution**

The dual problem is as below

$$\begin{aligned} \text{Maximize } Z_W &= 3w_1 + 5w_2 \\ \text{s.t.} \quad & 2w_1 + 8w_2 \geq 40 \\ & 3w_1 + 4w_2 \geq 50 \\ & w_1, w_2 \geq 0 \end{aligned}$$

To solve this,

$$\begin{aligned} \text{Maximize } Z'_W &= -3w_1 - 5w_2 \\ \text{s.t.} \quad & -2w_1 - 8w_2 + S_1 = 40 \\ & -3w_1 - 4w_2 + S_2 = -50 \\ & w_1, w_2, S_1, S_2 \geq 0 \end{aligned}$$

We solve this by *dual simplex method* (DSM). (refer Sect. 3.6 for DSM.)

**Simplex Table 1**

| | | $C_j \rightarrow$ | -3 | -5 | 0 | 0 | |
|---|---|---|---|---|---|---|---|
| Basic Variable (B.V.) | $C_B$ | $X_B$ | $w_1$ | $w_2$ | $S_1$ | $S_2$ | Min. ratio ($X_B/X_k$) |
| $S_1$ | 0 | -40 | -2 | -8 | 1 | 0 | 0/-40 |
| $S_2$ | 0 | -50 | [-3] | -4 | 0 | 1 | 0/-50← |
| | $w = 0$ | | 3 ↑ | 5 | 0 | 0 | ← $\Delta_j$ |

**Simplex Table 2**

| | | $C_j \rightarrow$ | -3 | -5 | 0 | 0 | |
|---|---|---|---|---|---|---|---|
| Basic Variable (B.V.) | $C_B$ | $X_B$ | $w_1$ | $w_2$ | $S_1$ | $S_2$ | Min. ratio ($X_B/X_k$) |
| $S_1$ | 0 | -20/3 | 0 | [-16/3] | 1 | -2/3 | 0/(-20/3)← |
| $w_1$ | -3 | 50/3 | 1 | 4/3 | 0 | -1/3 | -3/(50/3) |
| | $w = -50$ | | 0 | +1 ↑ | 0 | 3 | ← $\Delta_j$ |

**Simplex Table 3**

| | | $C_j \rightarrow$ | -3 | -5 | 0 | 0 | |
|---|---|---|---|---|---|---|---|
| Basic Variable (B.V.) | $C_B$ | $X_B$ | $w_1$ | $w_2$ | $S_1$ | $S_2$ | Min. ratio ($X_B/X_k$) |
| $w_2$ | -5 | 5/4 | 0 | 1 | -3/16 | 1/8 | – |
| $w_1$ | -3 | 15 | 1 | 0 | ¼ | -1/2 | – |
| | $w = -205/4$ | | 0 | 0 | 3/16 | 7/8 | ← $\Delta_j$ |

Therefore,

$$S_1 = 3/16, S_2 = 7/8, w = 205/4$$

For the values of primal problem, replace $S_1$, $S_2$ by $X_1$, $X_2$. Therefore, the values are $X_1 = 3/16$, $X_2 = 7/8$, $Z = 205/4$.

**Illustration 3.17**

Solve the following LP by DSM:

$$\begin{aligned} \text{Maximize } Z &= X_1 + 2X_2 + 3X_3 \\ \text{s.t.} \quad & 2X_1 - X_2 + X_3 \geq 4 \\ & X_1 + X_2 + 2X_3 \leq 8 \\ & X_2 - X_3 \geq 2 \\ & X_1, X_2, X_3 \geq 0 \end{aligned}$$

**Solution**

We convert the objective function to maximization type.

$$\begin{aligned} \text{Maximize } Z' &= -X_1 - 2X_2 - 3X_3 \\ \text{s.t.} \quad & -2X_1 + X_2 - X_3 + S_1 = -4 \\ & X_1 + X_2 + 2X_3 + S_2 = 8 \\ & -X_2 + X_3 + S_3 = -2 \\ & X_1, X_2, X_3, S_1, S_2, S_3 \geq 0 \end{aligned}$$

**Simplex Table 1**

| | | $C_j$ → | -1 | -2 | -3 | 0 | 0 | 0 | |
|---|---|---|---|---|---|---|---|---|---|
| Basic Variable (B.V.) | $C_B$ | $X_B$ | $X_1$ | $X_2$ | $X_3$ | $S_1$ | $S_2$ | $S_3$ | |
| $S_1$ | 0 | -4 | [-2] | 1 | -1 | 1 | 0 | 0 | ← |
| $S_2$ | 0 | 8 | 1 | 1 | 2 | 0 | 1 | 0 | |
| $S_3$ | 0 | -2 | 0 | -1 | 1 | 0 | 0 | 1 | |
| | $Z'_j = 0$ | | 1 | 2 | 3 | 0 | 0 | 0 | $\Delta_j$ ← |
| | | | ↑ | | | | | | |

There is a slight difference in DSM calculations from the normal simplex calculations as already discussed in the preceding sections. In DSM, first we find out the leaving variable and then the entering variable.

**To find out the leaving variable**. Inspect the $X_B$ column and choose the most negative value in that column. Here, it is Min [−4, −2], which is −4. So, mark row $S_1$ with (←), which is the leaving vector/variable/element.

**To find out the entering variable**. Here, we take the ratios of values in $\Delta_j$ row and pivotal row and choose the maximum value.

Ratios should be calculated against only negative values in the pivotal row, or leaving vector row. Here, it is

$$\text{Max}\left[\frac{+1}{-2}, \frac{+3}{-1}\right] = -\frac{1}{2}$$

Hence, $X_1$ is the entering element that replaces $S_1$.

**Simplex Table 2**

| | | $C_j$ → | -1 | -2 | -3 | 0 | 0 | 0 | |
|---|---|---|---|---|---|---|---|---|---|
| Basic Variable (B.V.) | $C_B$ | $X_B$ | $X_1$ | $X_2$ | $X_3$ | $S_1$ | $S_2$ | $S_3$ | |
| $X_1$ | -1 | 2 | +1 | -1/2 | 1/2 | -1/2 | 0 | 0 | |
| $S_2$ | 0 | 6 | 0 | 3/2 | 3/2 | 1/2 | 1 | 0 | |
| $S_3$ | 0 | -2 | 0 | -1 | 1 | 0 | 0 | 1 | ← |
| | $Z'_j = -2$ | | 0 | (5/2)/-2 | 5/2 | 1/2 | 0 | 0 | ← $\Delta_j$ |
| | | | | ↑ | | | | | |

**To find the leaving variable**. Since the only negative value is $-2$ in the $X_B$ column, i.e., in the $S_3$ row, $S_3$ is the leaving element.

For the entering element

$$\text{Take the ratio Min}\left[\frac{\Delta_2}{X_{32}}\right] = -\frac{5}{2}$$

Hence, $X_2$ is the entering element that replaces $S_1$.

**Simplex Table 3**

| | | $C_j$ → | -1 | -2 | -3 | 0 | 0 | 0 | |
|---|---|---|---|---|---|---|---|---|---|
| Basic Variable (B.V.) | $C_B$ | $X_B$ | $X_1$ | $X_2$ | $X_3$ | $S_1$ | $S_2$ | $S_3$ | |
| $X_1$ | -1 | 3 | 1 | 0 | 0 | -1/2 | 0 | -1/2 | |
| $S_1$ | 0 | 3 | 0 | 0 | 3 | ½ | 1 | 3/2 | |
| $X_3$ | -2 | 2 | 0 | 1 | -1 | 0 | 0 | 1 | |
| | $Z'_j = -7$ | | 0 | 0 | 5 | 1/2 | 0 | 5/2 | ← $\Delta_j$ |

Hence, the optimal solution is $Z = 7$, $X_1 = 3$, $X_3 = 2$.

### Illustration 3.18

Use dual simplex method to solve the following LP:

$$\begin{aligned} \text{Maximize } Z = \ & 3X_1 + X_2 \\ \text{s.t.} \quad & X_1 +_2 \geq 1 \\ & 2X_1 + 3X_2 \geq 2 \\ & X_1, X_2 \geq 0 \end{aligned}$$

### Solution

We write the problem as below

$$\begin{aligned} \text{Maximize } Z' = & -3X_1 - X_2 \\ \text{s.t.} \quad & -X_1 - X_2 + S_1 = 1 \\ & -2X_1 - 3X_2 + S_2 = -2 \\ & X_1, X_2, S_1, S_2 \geq 0 \end{aligned}$$

**Simplex Table 1**

| | | $C_j$ → | -3 | -1 | 0 | 0 | |
|---|---|---|---|---|---|---|---|
| Basic Variable (B.V.) | $C_B$ | $X_B$ | $X_1$ | $X_2$ | $S_1$ | $S_2$ | |
| $S_1$ | 0 | -1 | -1 | -1 | 1 | 0 | |
| $S_2$ | 0 | -2 | -2 | [-3] | 0 | 1 | ← |
| | $Z_j = 0$ | | 3 | 1 | 0 | 0 | $\Delta_j$ ← |
| | | | | ↑ | | | |

The leaving variable = Min $(-1, -2) = -2 \rightarrow S_2$ row.
The entering variable = Max $[(-3/2), (-1/3)] = -1/3 \rightarrow X_2$

**Simplex Table 2**

| | | $C_j$ → | -3 | -1 | 0 | 0 | |
|---|---|---|---|---|---|---|---|
| Basic Variable (B.V.) | $C_B$ | $X_B$ | $X_1$ | $X_2$ | $S_1$ | $S_2$ | |
| $S_1$ | 0 | -1/3 | -1/3 | 0 | 1 | [-1/3] | ← |
| $X_1$ | -1 | 2/3 | 2/3 | 1 | 0 | -1/3 | |
| | $Z'_j = -2/3$ | | 7/3 | 0 | 0 | 1/3 | $\Delta_j$ ← |
| | | | | | | ↑ | |

The leaving variable = Min $(-1/3) = -1/3$ as $X_2$ value under $X_B$ column in nonnegative, i.e., $X_{B2}$ is nonnegative.

$$\text{The entering variable} = \text{ Max}\left[\frac{7/3}{-1/3}, \frac{1/3}{-1/3}\right] = -1/1 \rightarrow S_2$$

So $S_2$ replaces $S_1$

**Simplex Table 3**

| | | $C_j$ → | -3 | -1 | 0 | 0 | |
|---|---|---|---|---|---|---|---|
| Basic Variable (B.V.) | $C_B$ | $X_B$ | $X_1$ | $X_2$ | $S_1$ | $S_2$ | |
| $S_2$ | 0 | 1 | 1 | 0 | -3 | 1 | |
| $X_2$ | -1 | 1 | 1 | 1 | -1 | 0 | |
| | $Z'_j = -1$ | | 2 | 0 | 1 | 0 | $\Delta_j$ ← |

As all $\Delta_j \geq 0$, the optimal values are

$$Z_j = 1, X_1 = 0, X_2 = 1$$

**Illustration 3.19**

Solve the following LP by DSM:

$$\begin{aligned} \text{Maximize } Z = \; & 6X_1 + 7X_2 + 3X_3 + 5X_4 \\ \text{s.t.} \quad & 5X_1 + 6X_2 - 3X_3 + 4X_4 \geq 12 \\ & X_2 + 5X_3 - 6X_4 \geq 10 \\ & 2X_1 + 5X_2 + X_3 + X_4 \geq 8 \\ & X_1, X_2, X_3, X_4 \geq 0 \end{aligned}$$

**Solution**

First, we convert the given problem into its standard primal form and add slack variables. Then, the problem becomes

$$\begin{aligned} \text{Maximize } Z' = \; & 6X_1 - 7X_2 - 3X_3 - 5X_4 \\ \text{s.t.} \quad & -5X_1 - 6X_2 + 3X_3 - 4X_4 \leq -12 \\ & -X_2 - 5X_3 + 6X_4 \leq -10 \\ & -2X_1 - 5X_2 - X_3 - X_4 \leq -8 \\ & X_1, X_2, X_3, X_4 \geq 0 \end{aligned}$$

Now, the inequalities in the constraints are converted into equalities.

$$\begin{aligned} -5X_1 - 6X_2 + 3X_3 - 4X_4 + S_1 &= -12 \\ -X_2 - 5X_3 + 6X_4 + S_2 &= -10 \\ -2X_1 - 5X_2 - X_3 - X_4 + S_3 &= -8 \end{aligned}$$

**Simplex Table 1**

| | | $C_j$ → | -6 | -7 | -3 | -5 | 0 | 0 | 0 | |
|---|---|---|---|---|---|---|---|---|---|---|
| Basic Variable (B.V.) | $C_B$ | $X_B$ | $X_1$ | $X_2$ | $X_3$ | $X_4$ | $S_1$ | $S_2$ | $S_3$ | |
| $S_1$ | 0 | -12 | -5 | -6 | 3 | -4 | 1 | 0 | 0 | ← |
| $S_2$ | 0 | -10 | 0 | -1 | -5 | 6 | 0 | 1 | 0 | |
| $S_3$ | 0 | -8 | -2 | -5 | -1 | -1 | 0 | 0 | 1 | |
| | $Z'_j = 0$ | | 6 | 7 | 3 | 5 | 0 | 0 | 0 | $\Delta_j$ ← |
| | | | | ↑ | | | | | | |

**Simplex Table 2**

| | | $C_j$ → | -6 | -7 | -3 | -5 | 0 | 0 | 0 | |
|---|---|---|---|---|---|---|---|---|---|---|
| Basic Variable (B.V.) | $C_B$ | $X_B$ | $X_1$ | $X_2$ | $X_3$ | $X_4$ | $S_1$ | $S_2$ | $S_3$ | |
| $X_2$ | -7 | 2 | 5/6 | 1 | -1/2 | 2/3 | -1/6 | 0 | 0 | |
| $S_2$ | 0 | -8 | 5/6 | 0 | -11/2 | 20/3 | -1/6 | 1 | 0 | ← |
| $S_3$ | 0 | 2 | 13/6 | 0 | -7/2 | 7/3 | -5/6 | 0 | 1 | |
| | $Z'_j = -14$ | | 1/6 | 0 | 13/2 | 1/3 | 7/6 | 0 | 0 | $\Delta_j$ ← |
| | | | | ↑ | | | | | | |

**Simplex Table 3**

| | | $C_j$ → | -6 | -7 | -3 | -5 | 0 | 0 | 0 | |
|---|---|---|---|---|---|---|---|---|---|---|
| Basic Variable (B.V.) | $C_B$ | $X_B$ | $X_1$ | $X_2$ | $X_3$ | $X_4$ | $S_1$ | $S_2$ | $S_3$ | |
| $X_2$ | -7 | 30/11 | 25/33 | 1 | 0 | 2/33 | -5/33 | -1/11 | 0 | |
| $X_3$ | -3 | 16/11 | -5/33 | 0 | 1 | -40/33 | 1/33 | -2/11 | 0 | ← |
| $S_3$ | 0 | 78/11 | 18/11 | 0 | 0 | -21/11 | -8/11 | -7/11 | 1 | |
| | $Z'_j = -258/11$ | | 38/33 | 0 | 0 | 271/33 | 32/33 | 13/11 | 0 | $\Delta_j$ ← |
| | | | | ↑ | | | | | | |

Now, the optimal solution can be written as

$$Z = 258/11; \quad X_i = 0; \quad X_2 = 30/11; \quad X_3 = 16/11; \quad X_4 = 0; \quad Z = 258/11$$

## 3.7 Special Cases of LP

The special cases of LP can be any of the following:

### 3.7.1 *Infeasible Solution*

If in the simplex method, the optimal simplex table has at least one artificial variable in the set of basic variables, then the LP has an infeasible solution space and hence an infeasible solution.

**Illustration 3.20**
Solve the following LP problem:

$$\begin{aligned} \text{Maximize } Z = \; & 10X_1 + 3X_2 \\ \text{s.t.} \quad & 2X_1 + 3X_2 \leq 18 \\ & 6X_1 + 5X_2 \geq 60 \\ & X_1, X_2 \geq 0 \end{aligned}$$

**Solution**
Introducing slack, surplus, and artificial variables, the problem becomes

$$\begin{aligned} \text{Maximize } Z = \; & 10X_1 + 3X_2 + 0.S_1 + 0.S_2 - MA_1 \\ \text{s.t.} \quad & 2X_1 + 3X_2 + S_1 = 18 \\ & 6X_1 + 5X_2 - S_2 + A_1 = 60 \\ & X_1, X_2, S_1, S_2, A_1 \geq 0 \end{aligned}$$

**Simplex Table 1**

| | | $C_j$ → | 10 | 3 | 0 | 0 | $-M$ | |
|---|---|---|---|---|---|---|---|---|
| Basic Variable (B.V.) | $C_B$ | $X_B$ | $X_1$ | $X_2$ | $S_1$ | $S_2$ | $A_1$ | Min. Ratio ($X_B/X_K$) |
| $S_1$ | 0 | 18 | [2] | 3 | 1 | 0 | 0 | 18/2 = 9← |
| $A_1$ | $-M$ | 60 | 16 | 5 | 0 | -1 | 1 | 60/6 = 10 |
| | $Z'_j = -60M$ | | $-6M-10$ ↑ | $-5M-3$ | 0 | $M$ | 0 | $\Delta_j$ ← |

**Simplex Table 2**

| | | $C_j$ → | 10 | 3 | 0 | 0 | -M |
|---|---|---|---|---|---|---|---|
| Basic Variable (B.V.) | $C_B$ | $X_B$ | $X_1$ | $X_2$ | $S_1$ | $S_2$ | $A_1$ |
| $X_1$ | 10 | 9 | [1] | 3/2 | 1/2 | 0 | 0 |
| $A_1$ | -M | 6 | 0 | -4 | -3 | -1 | 1 |
| | $Z'_j = -6M+90$ | | 0 | 12+4M | 5+3M | M | 0 |
| | | | ↑ | | | | |

Since all $\Delta_j \geq 0$, the solution obtained is optimal. But we observe that the value of $A_i$ is 6 in the $X_B$ column, which indicates that the problem has no feasible or infeasible solution.

#### 3.7.1.1 Graphical Representation of Infeasible Solution

**Illustration 3.21**

Solve the following LP graphically:

$$\begin{aligned} \text{Maximize } Z = \; & 6X_1 + 4X_2 \\ \text{s.t.} \quad & 4X_1 + 2X_2 \leq 4 \\ & 9X_1 + 12X_2 \geq 36 \\ & X_1, X_2 \geq 0 \end{aligned}$$

**Solution**

From the equation, we get

$$\begin{aligned} &4X_1 + 2X_2 = 4, \quad \text{which gives} \\ &X_1 = 1, \quad \text{when} \quad X_2 = 0 \\ &X_2 = 2, \quad \text{when} \quad X_1 = 0 \end{aligned}$$

The second constraint equation gives

$$\begin{aligned} &9X_1 + 12X_2 = 36 \\ &X_1 = 4, \quad \text{when } X_2 = 0 \\ &X_2 = 3, \quad \text{when } X_1 = 0 \end{aligned}$$

From the graph (Fig. 3.2), we see that there is no intersection of lines or equations. The shaded regions represent the feasible regions for each constraint equations. Therefore, we say that the given problem has no feasible or infeasible solution.

**Fig. 3.2** Graphical solution for $Z = 6X_1 + 4X_2$

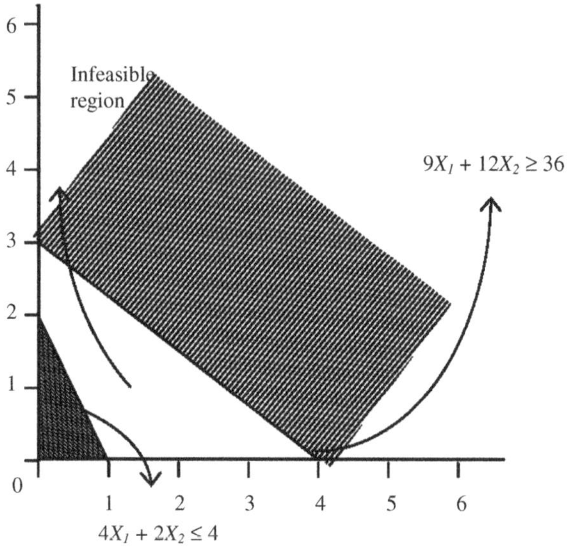


## *3.7.2 Unbounded Solution*

If in the simplex table, the constraint coefficients of the entering variables are $\leq 0$, then the solution space for the problem is unbounded, and hence no finite solution is possible.

**Illustration 3.22**
Solve the following LP:

$$\begin{aligned} \text{Maximize } Z = \; & 107X_1 + X_2 + 2X_3 \\ \text{s.t.} \quad & 14X_1 + X_2 - 6X_3 + 3X_4 = 7 \\ & 16X_1 + X_2 - 6X_3 \leq 5 \\ & 3X_1 - X_2 - X_3 \leq 0 \\ & X_1, X_2, X_3, X_4 \geq 0 \end{aligned}$$

**Solution**
The first constraint possesses an extra variable $X_4$, which does not appear in any of the constraints. Therefore, we divide the first constraint by the coefficient of $X_4$, i.e., 3, and treat $X_4$ as a slack variable. So, the problem becomes

$$\begin{aligned} \text{Maximize } Z = \; & 107X_1 + X_2 + 2X_3 + 0.S_1 + 0.S_2 + 0.X_4 \\ \text{s.t.} \quad & (14/3)X_1 + (1/3)X_2 - 2X_3 + X_4 = 7/3 \\ & 16X_1 + X_2 - 6X_3 + S_1 = 5 \\ & 3X_1 - X_2 - X_3 + S_2 = 0 \\ & X_1, X_2, X_3, S_1, S_2, X_4 \geq 0 \end{aligned}$$

**Simplex Table 1**

| Basic Variable (B.V.) | $C_B$ | $X_B$ | $X_1$ | $X_2$ | $X_3$ | $X_4$ | $S_1$ | $S_2$ | Min ratio ($X_B/X_K$) |
|---|---|---|---|---|---|---|---|---|---|
| | $C_j$ → | | 107 | 1 | 2 | 0 | 0 | 0 | |
| $X_4$ | 0 | 7/3 | 14/3 | 1/3 | -2 | 1 | 0 | 0 | 7/14=1/2 |
| $S_1$ | 0 | 5 | 16 | 1/2 | -6 | 0 | 1 | 0 | 5/16 |
| $S_2$ | 0 | 0 | [3] | -1 | -1 | 0 | 0 | 1 | 0/3← |
| | $Z_j = 0$ | | -107 | -1 | -2 | 0 | 0 | 0 | $\Delta_j$ ← |
| | | | ↑ | | | | | | |

**Simplex Table 2**

| Basic Variable (B.V.) | $C_B$ | $X_B$ | $X_1$ | $X_2$ | $X_3$ | $X_4$ | $S_1$ | $S_2$ | Min ratio ($X_B/X_K$) |
|---|---|---|---|---|---|---|---|---|---|
| | $C_j$ → | | 107 | 1 | 2 | 0 | 0 | 0 | |
| $X_4$ | 0 | 7/3 | 0 | 17/9 | -4/9 | 1 | 0 | -14/9 | – |
| $S_1$ | 0 | 5 | 0 | 35/6 | -2/3 | 0 | 1 | -16/3 | – |
| $X_1$ | 107 | 0 | 1 | -1/3 | -1/3 | 0 | 0 | 1/3 | – |
| | $Z_j = 0$ | | 0 | -110/3 | -113/3 | 0 | 0 | 107/3 | $\Delta_j$ ← |

The above solution is not optimal. The most negative value is present in the $X_3$ column, but $X_3$ cannot be the entering element as all elements under it are negative. Hence, we conclude that the above problem has an unbounded solution.

#### 3.7.2.1 Graphical Representation of Unbounded Solution

**Illustration 3.23**

Solve the following LP graphically:

$$\begin{aligned} \text{Maximize } Z = \; & 24X_1 + 50X_2 \\ \text{s.t.} \quad & 36X_1 + 9X_2 \geq 108 \\ & (15/2)X_1 - (5/2)X_2 \leq 15 \\ & X_1, X_2 \geq 0 \end{aligned}$$

**Solution**

We compute the coordinates by taking the constraints in equation form.
From the first constraint

$$36X_1 + 9X_2 = 108, \quad \text{we get}$$
$$X_1 = 3, \quad \text{when} \quad X_2 = 0$$
$$X_2 = 12, \quad \text{when} \quad X_1 = 0$$

From the second constraint

$$(15/2)X_1 - (5/2)X_2 = 15 \quad \text{we get}$$
$$X_1 = 2, \quad \text{when} \quad X_2 = 0$$
$$X_2 = -6, \quad \text{when} \quad X_1 = 0$$

It is clear from the graph (Fig. 3.3) that one side of the shaded space remains unclosed. Therefore, we conclude that the given problem has unbounded solution.

### *3.7.3 Unbounded Solution Space with Finite Solution*

**Illustration 3.24**
Solve the following LP problem:

$$\text{Maximize } Z = 12X_1 - 4X_2$$
$$\text{s.t.} \quad 4X_1 - 2X_2 \leq 108$$
$$2X_1 \leq 8$$
$$X_1, X_2 \geq 0$$

**Solution**
The given problem can be written as

**Simplex Table 1**

| | | $C_j \rightarrow$ | 12 | -4 | 0 | 0 | |
|---|---|---|---|---|---|---|---|
| Basic Variable (B.V.) | $C_B$ | $X_B$ | $X_1$ | $X_2$ | $S_1$ | $S_2$ | Min ratio ($X_B/X_K$) |
| $S_1$ | 0 | 4 | [4] | -2 | 1 | 0 | 4/4 = 1← |
| $S_2$ | 0 | 8 | 2 | 0 | 0 | 1 | 8/2 = 4 |
| | $Z_j = 0$ | | -12 ↑ | 4 | 0 | 0 | $\Delta_j$ ← |

**Simplex Table 2**

| | | $C_j \rightarrow$ | 12 | -4 | 0 | 0 | |
|---|---|---|---|---|---|---|---|
| Basic Variable (B.V.) | $C_B$ | $X_B$ | $X_1$ | $X_2$ | $S_1$ | $S_2$ | Min ratio ($X_B/X_K$) |
| $X_1$ | 12 | 1 | 1 | -1/2 | 1/4 | 0 | – |
| $S_2$ | 0 | 6 | 0 | [1] | -1/2 | 1 | 6/6 = 1← |
| | $Z_j = 12$ | | 0 | -2 ↑ | 3 | 0 | $\Delta_j$ ← |

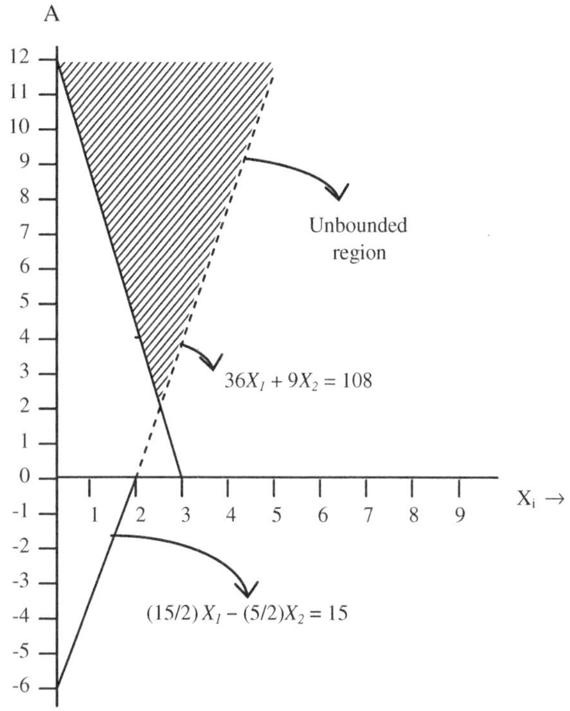


**Fig. 3.3** Graphical solution for $Z = 24X_1 + 50X_2$

**Simplex Table 3**

| | | $C_j \rightarrow$ | 12 | -4 | 0 | 0 | |
|---|---|---|---|---|---|---|---|
| Basic Variable (B.V.) | $C_B$ | $X_B$ | $X_1$ | $X_2$ | $S_1$ | $S_2$ | Min ratio ($X_B/X_K$) |
| $X_1$ | 12 | 4 | 1 | 0 | 0 | 1/2 | – |
| $X_2$ | -4 | 6 | 0 | 1 | -1/2 | 1 | – |
| | $Z_j = 24$ | | 0 | 0 | 2 | 2 | $\Delta_j \leftarrow$ |

Since all $\Delta_j \geq 0$, we have obtained the optimal solution with values $X_1 = 4$, $X_2 = 6$. But we observe that the $X_2$ column values in Simplex Table 1 have either negative or zero values (−2, 0). Hence, we say that the problem has optimal solution but does not possess a bounded feasible region.

#### 3.7.3.1 Graphical Representation of Unbounded Solution Space with Feasible Solution

**Illustration 3.25**
Solve the LP problem graphically:

$$\begin{aligned} \text{Maximize } Z = \; & 10X_1 - 4X_2 \\ \text{s.t.} \quad & 3X_1 \leq 108 \\ & -3X_1 + 6X_2 \geq 12 \\ & X_1, X_2 \geq 0 \end{aligned}$$

**Solution**
By taking the constraints into equation forms, we get the coordinates.
From the first constraint

$$3X_1 = 6, \quad \text{we get} \quad X_1 = 2$$

From the second constraint

$$\begin{aligned} -3X_1 + 6X_2 = 12 \quad & \text{we get} \\ X_1 = -4, \quad & \text{when} \quad X_2 = 0 \\ X_2 = 2, \quad & \text{when} \quad X_1 = 0 \end{aligned}$$

We see from the graph (Fig. 3.4) that element $X_2$ has unlimited value, but element $X_1$ has unlimited value. As the value of $X_2$ is less than zero in the objective function, it contributes to decrease in the objective function value on increasing $X_2$ and keeping $X_1$ constant. Hence, we say the problem has unbounded solution with finite solution.

The values will be $Z = 12$, $X_1 = 2$, $X_2 = 3$.

### *3.7.4 Alternate Optimal/Multiple Optimal Solution*

If in the simplex table the value of $C_j - Z_j$ for at least one non-basic variable is equal to 0, then the problem has alternate optimal solution.

**Illustration 3.26**
Solve the following LP problem:

$$\begin{aligned} \text{Maximize } Z = \; & 6X_1 + 4X_2 \\ \text{s.t.} \quad & 2X_1 + 3X_1 \leq 30 \\ & 3X_1 + 2X_2 \leq 24 \\ & X_1 + X_2 \geq 3 \\ & X_1, X_2 \geq 0 \end{aligned}$$

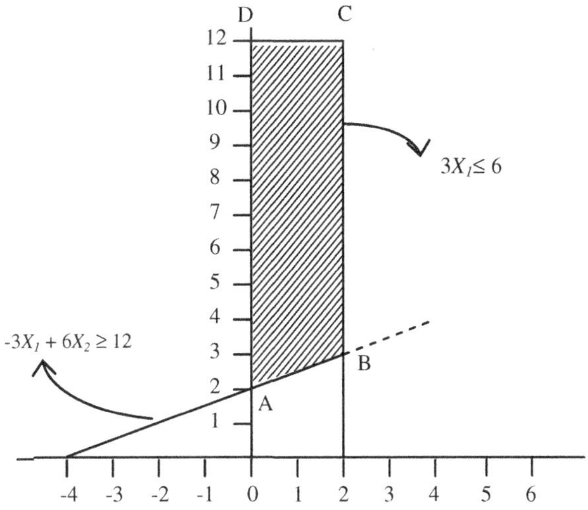


**Fig. 3.4** Graphical solution for $Z = 10X_1 - 4X_2$

State whether the solution is unique. Find the alternate solution on uniqueness not found to be existing.

### Solution

We rewrite the problem as

$$
\begin{aligned}
\text{Maximize } Z = \ & 6X_1 + 4X_2 + 0.S_1 + 0.S_2 + 0.S_3 - MA_1 \\
\text{s.t.} \quad & 2X_1 + 3X_1 + S_1 = 30 \\
& 3X_1 + 2X_2 + S_2 = 24 \\
& X_1 + X_2 + A_1 - S_3 = 3 \\
& X_1, X_2, S_1, S_2, S_3, A_1 \geq 0
\end{aligned}
$$

**Simplex Table 1**

| | | $C_j$ → | 6 | 4 | 0 | 0 | 0 | $-M$ | |
|---|---|---|---|---|---|---|---|---|---|
| Basic Variable (B.V.) | $C_B$ | $X_B$ | $X_1$ | $X_2$ | $S_1$ | $S_2$ | $S_3$ | $A_1$ | Min ratio $(X_B/X_K)$ |
| $S_1$ | 0 | 30 | 2 | 3 | 1 | 0 | 0 | 0 | 30/2 = 15 |
| $S_2$ | 0 | 24 | 3 | 2 | 0 | 1 | 0 | 0 | 24/3 = 8 |
| $A_1$ | $-M$ | 3 | [1] | 1 | 0 | 0 | $-1$ | 1 | 3/1 = 3← |
| | $Z_j = -3M$ | | $-M-6$ | $-M-4$ | 0 | 0 | $M$ | 0 | $\Delta_j$ ← |
| | | | ↑ | | | | | | |

**Simplex Table 2**

| Basic Variable (B.V.) | $C_B$ | $X_B$ | $X_1$ | $X_2$ | $S_1$ | $S_2$ | $S_3$ | $A_1$ | Min ratio ($X_B/X_K$) |
|---|---|---|---|---|---|---|---|---|---|
| | $C_j$ → | | 6 | 4 | 0 | 0 | 0 | -$M$ | |
| $S_1$ | 0 | 24 | 0 | 1 | 1 | 0 | 2 | -2 | 24/2 = 12 |
| $S_2$ | 0 | 15 | 0 | -1 | 0 | 1 | [3] | -3 | 15/3 = 5← |
| $X_1$ | 6 | 3 | 1 | 1 | 0 | 0 | -1 | 1 | – |
| | $Z_j$ = 18 | | 0 | 1 | 0 | 0 | -6 | 6 +$M$ | $\Delta_j$ ← |
| | | | | | | | ↑ | | |

**Simplex Table 3**

| Basic Variable (B.V.) | $C_B$ | $X_B$ | $X_1$ | $X_2$ | $S_1$ | $S_2$ | $S_3$ | $A_1$ | Min ratio ($X_B/X_K$) |
|---|---|---|---|---|---|---|---|---|---|
| | $C_j$ → | | 6 | 4 | 0 | 0 | 0 | -$M$ | |
| $S_1$ | 0 | 14 | 0 | [5/3] | 1 | -2/3 | 0 | 0 | 14/(5/3) = 42/5← |
| $S_3$ | 0 | 5 | 0 | -1/3 | 0 | 1/3 | 1 | -1 | – |
| $X_1$ | 6 | 8 | 1 | 2/3 | 0 | 1/3 | 0 | 0 | 24/2 = 12 |
| | $Z_j$ = 48 | | 0 | 0 | 0 | 2 | 0 | $M$ | $\Delta_j$ ← |
| | | | | ↑ | | | | | |

The values are $Z_j = 48, X_1 = 8, X_2 = 0$.

In Simplex Table 3, we notice that $X_2$ is not entering the BV column and also has zero value in the $\Delta_j$ row. Hence, we say that there is an alternate solution, and the above solution is not optimal.

(*Note* After reaching optimality, if at least one of the non-basic or decision variables possesses a zero value in $\Delta_j$ row, then alternative optimal solution exists. If not, it is called a unique solution.)

We find the alternate solution by selecting $X_2$ as the entering element, which replaces $S_1$.

**Simplex Table 4**

| Basic Variable (B.V.) | $C_B$ | $X_B$ | $X_1$ | $X_2$ | $S_1$ | $S_2$ | $S_3$ | $A_1$ | |
|---|---|---|---|---|---|---|---|---|---|
| | $C_j$ → | | 6 | 4 | 0 | 0 | 0 | -$M$ | |
| $X_2$ | 4 | 42/5 | 0 | 1 | 3/5 | -2/5 | 0 | 0 | |
| $S_3$ | 0 | 39/5 | 0 | 0 | 1/5 | 1/5 | 1 | -1 | |
| $X_1$ | 6 | 12/5 | 1 | 0 | -2/5 | 3/5 | 0 | 0 | |
| | $Z_j$ = 48 | | 0 | 0 | 0 | 2 | 0 | $M$ | $\Delta_j$ ← |

Now, the solution obtained is an alternate optimal solution. The values are

$$Z_j = 48 \quad X_1 = 12/5 \quad X_2 = 42/5$$

### 3.7.5 Degeneracy

If in any one iteration there is a tie for the minimum ratio to maintain feasibility, the objective function of the nest iteration will be the same as that of the current iteration. This will occur when in any of the corner points of the feasible region, more than two lines pass through. This will result in carrying out more iteration without any improvement in the value of the objective function.

**Illustration 3.27**
Solve the following LP problem:

$$\begin{aligned} \text{Maximize } Z = \; & 6X_1 + 18X_2 \\ \text{s.t.} \quad & 2X_1 + 8X_1 \leq 16 \\ & 2X_1 + 4X_2 \leq 8 \\ & X_1, X_2 \geq 0 \end{aligned}$$

**Solution**
We introduce slack variables $S_1$ and $S_2$.

$$\begin{aligned} \text{Maximize } Z = \; & 6X_1 + 18X_2 + 0.S_1 + 0.S_2 \\ \text{s.t.} \quad & 2X_1 + 8X_1 + S_1 = 16 \\ & 2X_1 + 4X_2 + S_2 = 8 \\ & X_1, X_2, S_1, S_2 \geq 0 \end{aligned}$$

**Simplex Table 1**

| | | $C_j \rightarrow$ | 6 | 18 | 0 | 0 | |
|---|---|---|---|---|---|---|---|
| Basic Variable (B.V.) | $C_B$ | $X_B$ | $X_1$ | $X_2$ | $S_1$ | $S_2$ | Min ratio ($X_B/X_K$) |
| $S_1$ | 0 | 16 | 2 | 8 | 1 | 0 | 16/8 = 2 |
| $S_2$ | 0 | 8 | 2 | [4] | 0 | 1 | 8/4 = 2 ← |
| | $Z_j = 0$ | | -6 | -18 ↑ | 0 | 0 | $\Delta_j$ ← |

From the minimum ratio column, we notice that both the ratio values are the same. This means there is a tie for the pivotal row, i.e., either $S_1$ or $S_2$ can leave the basis which will be replaced by $X_2$. This situation is termed as *degeneracy*. In order to zero in on the pivotal row, we follow a systematic procedure.

1. Identify the entering variable column.
2. Take the ratios of elements in unit matrix columns with the corresponding elements in the entering variable column and take the minimum ratio.

In this problem, there are two slack variable columns $S_1$ and $S_2$, which form the unit matrix. First, we take the ratios under $S_1$. If there is a tie here also with respect to ratio values, then switch to the next, i.e., $S_2$ column.
The ratio values will be [(1/8), (0/4)].
We have to take Min [(1/8), (0/4)] which is (0/4) = 0.
Therefore, $S_2$ is the leaving element with 4 as the pivot element.

3. If artificial variables are present in the BV column and a situation of degeneracy is experienced, it is preferred to take the artificial variable as the leaving element. In such cases, there is no need to calculate ratios that involve elements from unit matrix. The normal procedure to resolve degeneracy will not be applicable in such cases.

Proceeding further,

**Simplex Table 2**

| | | $C_j$ → | 6 | 18 | 0 | 0 | |
|---|---|---|---|---|---|---|---|
| Basic Variable (B.V.) | $C_B$ | $X_B$ | $X_1$ | $X_2$ | $S_1$ | $S_2$ | |
| $S_1$ | 0 | 0 | -2 | 0 | 1 | -2 | |
| $X_2$ | 18 | 2 | 1/2 | 1 | 0 | 1/4 | |
| | $Z_j = 36$ | | 3 | 0 | 0 | 7/2 | $\Delta_j$ ← |

Hence, the optimal solution is

$$Z = 36 \quad X_1 = 0 \quad X_2 = 2$$

**Illustration 3.28**

Solve the following LP:

$$\begin{aligned} \text{Maximize } Z = \; & 5X_1 - 2X_2 + 3X_3 \\ \text{s.t.} \quad & 2X_1 + 2X_2 - X_3 \geq 2 \\ & 3X_1 - 4X_2 \leq 3 \\ & X_2 + 3X_3 \leq 5 \\ & X_1, X_2, X_3 \geq 0 \end{aligned}$$

**Solution**

We add slack, surplus, and artificial variances, thereby the problem becomes

$$\begin{aligned} \text{Maximize } Z = \; & 5X_1 - 2X_2 + 3X_3 + 0.S_1 + 0.S_2 + 0.S_3 = -MA_1 \\ \text{s.t.} \quad & 2X_1 + 2X_2 - X_3 - S_1 + A_1 = 2 \\ & 3X_1 - 4X_2 + S_2 = 3 \\ & X_2 + 3X_3 + S_3 = 5 \\ & X_1, X_2, X_3, S_1, S_2, S_3, A_1 \geq 0 \end{aligned}$$

**Simplex Table 1**

| | $C_j$ → | | 5 | -2 | 3 | 0 | 0 | 0 | $-M$ | |
|---|---|---|---|---|---|---|---|---|---|---|
| Basic Variable (B.V.) | $C_B$ | $X_B$ | $X_1$ | $X_2$ | $X_3$ | $S_1$ | $S_2$ | $S_3$ | $A_1$ | Min ratio ($X_B/X_K$) |
| $A_1$ | $-M$ | 2 | [2] | 2 | 1 | -1 | 0 | 0 | 1 | 2/2 = 1← |
| $S_2$ | 0 | 3 | 3 | -4 | 0 | 0 | 1 | 0 | 0 | 3/3 = 1 |
| $S_3$ | 0 | 5 | 0 | 1 | 3 | 0 | 0 | 1 | 0 | – |
| | $Z_j = -2M$ | | $-2M-5$ | $-2M+2$ | $M-3$ | $M$ | 0 | 0 | 0 | ← $\Delta_j$ |
| | | | ↑ | | | | | | | |

Since an artificial variable is present in BV column, it is directly taken as the leaving element without applying the procedure to resolve degeneracy.

**Simplex Table 2**

| | $C_j$ → | | 5 | -2 | 3 | 0 | 0 | 0 | $-M$ | |
|---|---|---|---|---|---|---|---|---|---|---|
| Basic Variable (B.V.) | $C_B$ | $X_B$ | $X_1$ | $X_2$ | $X_3$ | $S_1$ | $S_2$ | $S_3$ | $A_1$ | Min ratio ($X_B/X_K$) |
| $X_1$ | 5 | 1 | [1] | 1 | 1/2 | -1/2 | 0 | 0 | 1/2 | – |
| $S_2$ | 0 | 0 | 0 | 7 | 3/2 | 3/2 | 1 | 0 | -3/2 | 0/(3/2)← |
| $S_3$ | 0 | 5 | 0 | 1 | 3 | 0 | 0 | 1 | 0 | 5/3 |
| | $Z_j = 5$ | | 0 | 7 | -11/2 | -5/2 | 0 | 0 | $M-(5/2)$ | ← $\Delta_j$ |
| | | | ↑ | | | | | | | |

**Simplex Table 3**

| | $C_j$ → | | 5 | -2 | 3 | 0 | 0 | 0 | $-M$ | |
|---|---|---|---|---|---|---|---|---|---|---|
| Basic Variable (B.V.) | $C_B$ | $X_B$ | $X_1$ | $X_2$ | $X_3$ | $S_1$ | $S_2$ | $S_3$ | $A_1$ | Min ratio ($X_B/X_K$) |
| $X_1$ | 5 | 1 | 1 | -4/3 | 0 | 0 | 1/3 | 0 | 1 | – |
| $X_3$ | 3 | 0 | 0 | -14/3 | 1 | 1 | 2/3 | 0 | -1 | – |
| $S_3$ | 0 | 5 | 0 | [15] | 0 | -3 | -2 | 1 | 3 | 5/15 = 1/3← |
| | $Z_j = 5$ | | 0 | -56/3 | 0 | 3 | 11/3 | 0 | $M+2$ | ← $\Delta_j$ |
| | | | | ↑ | | | | | | |

**Simplex Table 4**

| | $C_j$ → | | 5 | -2 | 3 | 0 | 0 | 0 | $-M$ | |
|---|---|---|---|---|---|---|---|---|---|---|
| Basic Variable (B.V.) | $C_B$ | $X_B$ | $X_1$ | $X_2$ | $X_3$ | $S_1$ | $S_2$ | $S_3$ | $A_1$ | Min ratio ($X_B/X_K$) |
| $X_1$ | 5 | 13/9 | 1 | 0 | 0 | -4/15 | 7/45 | 4/45 | 19/15 | – |
| $X_3$ | 3 | 14/9 | 0 | 0 | 1 | [1/15] | 2/45 | 14/45 | -1/15 | 70/3← |
| $X_2$ | -2 | 1/3 | 0 | 1 | 0 | -15 | -2/15 | 1/15 | 1/5 | – |
| | $Z_j = 101/9$ | | 0 | 0 | 0 | -11/15 | 53/45 | 56/45 | $M+86/15$ | ← $\Delta_j$ |
| | | | | | | ↑ | | | | |

**Simplex Table 5**

| | $C_j$ → | | 5 | -2 | 3 | 0 | 0 | 0 | -$M$ | |
|---|---|---|---|---|---|---|---|---|---|---|
| Basic Variable (B.V.) | $C_B$ | $X_B$ | $X_1$ | $X_2$ | $X_3$ | $S_1$ | $S_2$ | $S_3$ | $A_1$ | Min ratio ($X_B/X_K$) |
| $X_1$ | 5 | 23/3 | 1 | 0 | 4 | 0 | 1/3 | 4/3 | 1 | – |
| $S_1$ | 0 | 70/3 | 0 | 0 | 15 | 1 | 2/3 | 14/3 | -1 | – |
| $X_2$ | -2 | 5 | 0 | 1 | 3 | 0 | 0 | 1 | -74/5 | – |
| | $Z_j$ = 85/3 | | 0 | 0 | 11 | 0 | 5/3 | 14/3 | $M$ + 173/5 | ← $\Delta_j$ |

It is evident that all $\Delta_j \geq 0$. Therefore, we conclude by writing the optimal solution as

$$Z = 85/3 \quad X_1 = 23/3 \quad X_2 = 5 \quad X_3 = 0$$

## 3.7.6 *Graphical Representation of Degeneracy*

**Illustration 3.29**

Use graphical method to solve the following LP:

$$\begin{aligned} \text{Maximize } Z = \; & 200X_1 + 100X_2 \\ \text{s.t.} \quad & 12X_1 + 18X_2 \leq 72 \\ & 2X_1 \leq 8 \\ & 2X_2 \leq 8/3 \\ & X_1, X_2 \geq 0 \end{aligned}$$

**Solution**

Taking the first constraint in equation form, we get

$$\begin{aligned} X_1 &= 6, \quad \text{when} \quad X_2 = 0 \\ X_2 &= 4, \quad \text{when} \quad X_1 = 0 \end{aligned}$$

From the second constraint, we get

$$2X_1 = 8, \quad \therefore X_1 = 4$$

Similarly, from the third constraint, we get

$$2X_2 = 8/3, \quad \therefore X_2 = 4/3$$

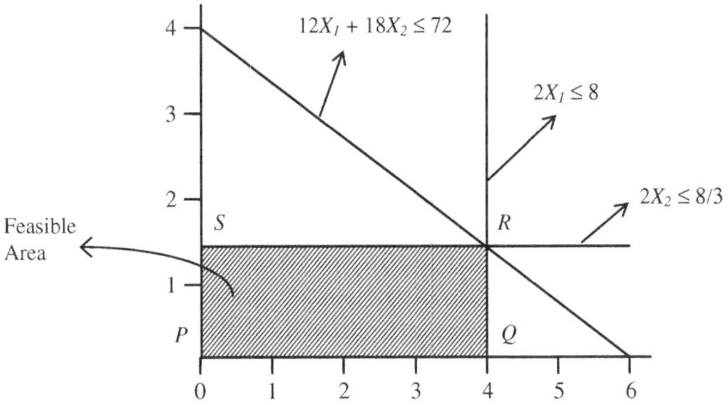


**Fig. 3.5** Graphical solution for $Z = 200X_1 + 100X_2$

Now, we plot these values (Fig. 3.5).
The corner point is $R$ with coordinates (4,4/3).
The values are computed as

$$Z(P) = 0$$
$$Z(Q) = 800$$
$$Z(R) = 2,800/3 = 933.34$$
$$Z(S) = 400/3 = 133.33$$

$Z$ value is maximum at $R$. Hence, the solution is

$$Z = 2,800/3 \quad X_1 = 4 \quad X_2 = 4/3$$

## 3.8 Review Questions

1. What are the assumptions of LPP?
2. Explain

   (a) Unbounded solution
   (b) Infeasible solution
   (c) Alternate optimum solution
   (d) Slack variable
   (e) Surplus variable
   (f) Degenerate solution
   (g) Basic variable
   (h) Interim variable

3. An organization produces two types of products, $X$ and $Y$, respectively. Each product has to pass through machines $M_1$ and $M_2$. The processing time for one unit of $X$ is 4 h on $M_1$ and 11 h on $M_2$. The maximum usage of the machines is Rs. 40 h and 60 h, respectively, in a week. The organization earns Rs. 6 and Rs. 8 as profit per unit selling $X$ and $Y$ respectively. Formulate the above as an LP model and obtain the production volume for each of the products that maximizes the total profits.
4. A firm manufactures headache pills in two sizes A and B. Size A contains 2 g of aspirin 5 g of bicarbonate and 1 g of codeine. Size B contains 1 g of aspirin, 8 g of bicarbonate and 6 g of codeine. It is found by users that it requires at least 12 g of aspirin, 74 g of bicarbonate and 24 g of codeine for providing immediate effect. It is required to determine the least number of pills a patient should take to get immediate relief. Formulate the problem as an LPP and determine the number of pills to be taken by the patient to get immediate relief.
5. Two alloys A and B are made from different metals I, II, III, IV according to the following specifications.

   A: at most 80 % of I, at most 30 % of II, at least 50 % of III.
   B: between 40 % and 60 % of II, at least 30 % of III, at most 70 % of IV.

   The four metals are extracted from three different ores whose constituents' percentage of these metals, maximum availability, and cost per ton are as follows.
   Assuming the selling prices of alloys A and B are Rs. 200 and Rs. 300 per ton, respectively, formulate the above as LPP. Selecting appropriate objective and constraint functions, find all the values on non-basic variables.

| Ore | Maximum quantity (tons) | I | II | III | IV | Others | Price (Rs. Per km) |
|---|---|---|---|---|---|---|---|
| 1 | 1000 | 20 | 10 | 30 | 30 | 10 | 30 |
| 2 | 2000 | 10 | 20 | 30 | 30 | 10 | 40 |
| 3 | 3000 | 5 | 5 | 70 | 20 | 0 | 50 |

6. A used car dealer wishes to stock up his lot to maximize his profit. He can select cars A, B, and C which are valued wholesale at Rs. 5,000, Rs. 7,000, and Rs. 8,000 respectively. These can be sold at Rs. 6,000, Rs. 8,500, and Rs. 10,500 respectively.

   For each car type, the probabilities of sale are

| Type of Car | A | B | C |
|---|---|---|---|
| Prob of sale in 90 days | 0.7 | 0.8 | 0.6 |

Formulate the above LPP. For every two cars of B, he should have one car of type A or C. If he has Rs. 1,00,000 to invest, what should he buy to maximize the expected gain?

7. A farmer has 1,000 acres of land on which he can grow corn, wheat, or soyabeans. Each acre of corn costs Rs. 100 for preparation, requires 7 man-days of work, and yields a profit of Rs. 30. An acre of wheat costs Rs. 120 to prepare, requires 10 man-days of work, and yields a profit of Rs. 40. An acre of soyabeans costs Rs. 70 to prepare, requires 8 man-days of work, and yields a profit of Rs. 20. If the farmer has Rs. 1,00,000 for preparation and can count on 8,000 man-days of work, how many acres should be allocated to each crop to maximize profit?
8. Solve the following Linear Programming Problem:

(a) Maximize $Z = x_1 - x_2 + 3x_3$
subject to

$$x_1 + x_2 + x_3 \leq 10$$
$$2x_1 - x_3 \leq 2$$
$$2x_1 - 2x_2 + 3x_3 \leq 0$$
$$\text{and } x_1, x_2, x_3 \geq 0$$

(b) Maximize $Z = x_1 + x_2 + 3x_3$
subject to

$$3x_1 + 2x_2 + x_3 \leq 3$$
$$2x_1 + x_2 + 2x_3 \leq 2$$
$$\text{and } x_1, x_2, x_3 \geq 0$$

9. Solve
Maximize $Z = 214x_1 + 2x_2 + 2x_3$
subject to

$$28x_1 + 2x_2 - 12x_3 + 6x_4 = 14$$
$$32x_1 + x_2 - 12x_3 \leq 10$$
$$6x_1 - 2x_2 - 2x_3 \leq 0$$
$$\text{and } x_1, x_2, x_3, x_4 \geq 0$$

(**Hint:** Treat $x_4$ as a slack variable and divide the first constraint equation by 6)

10. A firm has an advertising budget of Rs. 7,20,000. It wishes to allocate this budget to two media: Magazines and Television, so that total expenditure is maximized. Each page of Magazine advertising is estimated to result in 60,000 exposures, whereas each spot on Television is estimated to result in 1,20,000

exposures. Each page of magazine advertising costs Rs. 9,000 and each spot on television costs Rs. 12,000. An additional condition is that the firm has specified that at least two pages of magazine advertising be used and at least 3 spots on television. Determine the optimum media mix for this firm.

11. Solve
Maximize $Z = 10x_1 - 4x_2 + 6x_3$
subject to

$$4x_1 + 4x_2 - 2x_3 \geq 4$$
$$6x_1 - 8x_2 \leq 6$$
$$2x_2 + 6x_3 \leq 10$$
$$\text{and } x_1, x_2, x_3, x_4 \geq 0$$

12. What is an artificial variable and why is it necessary to introduce it?
13. Describe the two phase process of solving an LPP by simplex method. Why is an artificial vector that leaves the basis once never considered again for re-entry into the basis?
14. A cabinet manufacturer produces 10,000 cabinets for TV sets, stereo systems, and radios, each of which must be assembled and rated. Each TV cabinet requires 3 h to assemble, 5 h to decorate and 0.1 h to crate and earns a profit of Rs. 10. Each stereo cabinet requires 10 h to assemble, 8 h to decorate, and 0.6 h to crate, and earns a profit of Rs. 25. Each radio cabinet requires 1 h to assemble, 1 h to decorate, and 0.1 h to crate, and earns a profit of Rs. 3. The manufacturer has a maximum of 30,000, 40,000 and 120 h available for assembling, decorating, and crating respectively.

(a) Formulate the above problem as an LP.
(b) Use simplex method to find out how many units of each product should be manufactured to maximize profit.
(c) Does the problem have a unique solution?

15. Solve
Maximize $Z = 5x_1 - 2x_2 + 3x_3$
subject to

$$2x_1 + 2x_2 - x_3 \geq 2$$
$$3x_1 - 4x_2 \leq 3$$
$$x_2 + 3x_3 \leq 5$$
$$\text{and } x_1, x_2, x_3 \geq 0$$

16. Solve the following LP:
    Maximize $Z = \frac{3}{4}x_1 - 20x_2 + \frac{1}{2}x_3 - 6x_4$
    subject to

$$\frac{1}{4}x_1 - 8x_2 - x_3 + 9x_4 \leq 0$$
$$\frac{1}{2}x_1 - 12x_2 - \frac{1}{2}x_3 + 3x_4 \leq 0$$
$$x_3 \leq 1$$
$$\text{and } x_1, x_2, x_3, x_4 \geq 0$$

(**Hint:** A case of degeneracy-degenerate optimal solution)

17. Obtain the dual of the following LP problem:
    Maximize $f(x) = 2x_1 + 5x_2 + 6x_3$
    subject to

$$5x_1 + 6x_2 - x_3 \leq 3$$
$$-x_1 + x_2 + 3x_3 \geq 4$$
$$7x_1 - 2x_2 - x_3 \leq 10$$
$$x_1 - 2x_2 + 5x_3 \geq 3$$
$$4x_1 + 7x_2 - 2x_3 \geq 2$$
$$\text{and } x_1, x_2, x_3 \geq 0$$

18. Obtain the dual of the following LPP:
    Maximize $Z = 6x_1 - 4x_2 + 8x_3$
    subject to

$$5x_1 + 10x_2 + 6x_3 \geq 12$$
$$8x_1 + 3x_2 + 5x_3 \geq 8$$
$$9x_1 - 5x_2 - x_3 \leq 10$$
$$2x_1 - 4x_2 + 5x_3 \geq 5$$
$$7x_1 + 10x_2 - 5x_3 \geq 5$$
$$\text{and } x_1, x_2, x_3 \geq 0$$

19. Solve
    Maximize $Z = x_1 + 2x_2 + 3x_3$
    subject to

$$x_1 + 2x_2 + 3x_3 \leq 10$$
$$x_1 + x_2 \leq 5$$
$$x_1 \leq 1$$
$$\text{and } x_1, x_2, x_3 \geq 0$$

(**Hint:** Alternative optimal solutions exist, 3 alternate solutions are possible)

20. Solve the following LPP:

Maximize $Z = 20x_1 - 10x_2 + x_3$
subject to

$$3x_1 - 3x_2 + 5x_3 \leq 50$$
$$x_1 + x_3 \leq 10$$
$$x_1 - x_2 + 4x_3 \leq 20$$
$$\text{and}\, x_1, x_2, x_3 \geq 0$$

By inspecting the constraints, determine the direction in which the solution space is unbounded.

21. GTTC produces three types of tools – $T_1$, $T_2$, and $T_3$. In the process, two raw materials are consumed. The other data are given below –

| Raw material | Number of units per tool | | |
|---|---|---|---|
| | $T_1$ | $T_2$ | $T_3$ |
| $M_1$ | 3 | 5 | 6 |
| $M_2$ | 5 | 3 | 4 |

The daily availability is 1,000 and 1,200 units respectively. The daily demand must at least be 500 units. Will the demand be satisfied? If not, what is the most GTTC can provide of the three tools?

# Chapter 4
# Transportation Models and Its Variants

The key takeaways for the reader from this chapter are as follows:

- Introduces transportation problem
- Discusses different types of transportation problems
- Different methods of solution
- Discusses transshipment and assignment problem.

## 4.1 Introduction

Transportation problem is a special case of LPP. In a typical transportation problem, the objective is to transport various amounts of a single homogeneous commodity that are initially stored at various origins, to different destinations in such a way that the total transportation cost is minimum.

## 4.2 Mathematical Formulation

Suppose there are M origins; $i$th origin has $A_i$ units of a certain product, there are $N$ destinations ($N$ may or may not be equal to $M$) with destination $j$ requiring $B_j$ units, and shipping costs to each of the $N$ destinations are known in terms of kilometers, shipping hours, etc. Let $C_{ij}$ be the cost of shipping one unit of a particular product from $i$th origin to $j$th destination and $X_{ij}$ be the amount to be shipped from the $i$th origin to the $j$th destination. The assumption is total availability ($\Sigma A_i$) satisfies the total requirements ($\Sigma B_i$), i.e.,

$$(\Sigma A_i) = (\Sigma B_i) \quad \forall i = 1, 2, \ldots\ldots M, \ \forall j = 1, 2, \ldots\ldots N$$

R. Srinivasan, *Strategic Business Decisions*,
DOI: 10.1007/978-81-322-1901-9_4, © Springer India 2014

We have to determine no-negativity values ($\geq 0$) of $X_{ij,}$ which satisfies both the availability and requirement constraints, i.e.,

$$\sum_{i=1}^{N} X_{ij} = A_i \quad \forall i = 1, 2, \ldots\ldots M$$

$$\sum_{i=1}^{M} X_{ij} = B_j \quad \forall i = 1, 2, \ldots\ldots N$$

The objective is to minimize the total cost of transportation (shipping), i.e.,

$$Z = \sum_{i=1}^{M} \sum_{i=1}^{N} X_{ij} C_{ij}$$

Since the objective function and the constraints are linear in $X_{ij,}$ the problem is a special case of LPP.

The assignment problem is a special case of transportation problem, where each origin is associated with **one** and **only one** destination, i.e., $M = N$. The numerical evaluation of such association is called "effectiveness" (instead of transportation costs). All $A_i$ and $B_j$ are units and $X_{ij}$ limited to either 0 or 1. Then, exactly, $N$ of the $X_{ij}$ can be nonzero (i.e., unity), one for each origin and destination.

## 4.3 Types of Transportation Problems

Transportation problems can be

1. Balanced transportation problem
2. Unbalanced transportation problem.

In case of a balanced transportation problem, the sum of the supplies of all sources is equal to the sum of demands of all destinations, i.e.,

$$\sum_{i=1}^{M} A_i = \sum_{j=1}^{N} B_j$$

In the case of unbalanced transportation problem, the above equality does not hold or is not there, i.e.,

$$\sum_{i=1}^{M} A_i \neq \sum_{j=1}^{N} B_{\mathrm{j}}$$

where $A_i$ can be supplies/origins/factories/requirements, and $B_j$ can be demand/destinations/warehouses/availability.

Normally in such cases, we can convert it into a balanced transportation problem by adding a source or destination as shown in the examples below.

**Illustration 4.1**

A case of excess requirement.

| | | Destination (To) | | | Availability |
|---|---|---|---|---|---|
| | | $D_1$ | $D_2$ | $D_3$ | ↓ |
| | $S_1$ | 20 | 40 | 10 | 250 |
| Source (From) | $S_2$ | 25 | 60 | 10 | 250 |
| | $S_3$ | 10 | 35 | 50 | 500 |
| Requirement | → | 250 | 350 | 250 | |

**Solution**

The sum of requirements $(\Sigma A_i) = 850$

The sum of availability $(\Sigma B_j) = 1{,}000$

$$\therefore \Sigma A_j \neq \Sigma B_j \quad \text{or} \quad \Sigma A_j < \Sigma B_j$$

In these cases, we add one more dummy destination with 0 values, with the excess requirement value. Therefore,

| | | Destination (To) | | | | Availability |
|---|---|---|---|---|---|---|
| | | $D_1$ | $D_2$ | $D_3$ | $D_4$ | ↓ |
| | $S_1$ | 20 | 40 | 10 | 0 | 250 |
| Source (From) | $S_2$ | 25 | 60 | 10 | 0 | 250 |
| | $S_3$ | 10 | 35 | 50 | 0 | 500 |
| Requirement | → | 250 | 350 | 250 | 150 | |

The excess requirement $= \Sigma B_j - \Sigma A_j = 1{,}000 - 850 = 150$

**Illustration 4.2**
A case of excess availability.

| | | Destination (To) | | | | Availability |
|---|---|---|---|---|---|---|
| | | $D_1$ | $D_2$ | $D_3$ | $D_4$ | ↓ |
| | $S_1$ | 10 | 6 | 6 | 10 | 250 |
| Source (From) | $S_2$ | 14 | 4 | 13 | 10 | 150 |
| | $S_3$ | 9 | 3 | 2 | 9 | 400 |
| Requirement | → | 150 | 150 | 400 | 200 | |

**Solution**
Here, $\Sigma A_i = 900$, $\Sigma B_j = 800$. We make out that $\Sigma A_i > \Sigma B_j$. In these cases, we add an extra dummy source to fulfill the availability with the excess availability with the excess availability value.

| | | Destination (To) | | | | Availability |
|---|---|---|---|---|---|---|
| | | $D_1$ | $D_2$ | $D_3$ | $D_4$ | ↓ |
| | $S_1$ | 10 | 6 | 6 | 10 | 250 |
| | $S_2$ | 14 | 4 | 13 | 10 | 150 |
| Source (From) | $S_3$ | 9 | 3 | 2 | 9 | 400 |
| | $S_4$ | 0 | 0 | 0 | 0 | 100 |
| Requirement | → | 150 | 150 | 400 | 200 | |

$$\text{The excess requirement} = \Sigma A_i - \Sigma B_j = 900 - 800 = 100$$

## 4.4 Solving the Transportation Problem

The first step in the approach toward solving a transportation problem is to find the initial basic feasible solution (IBFS) and further optimize the IBFS to obtain an optimal solution. There are five methods or approaches to solve a transportation problem. They are discussed below.

### 4.4.1 Methods to Solve a Transportation Problem

Five methods can be used to solve the transportation problem. They are as below:

1. North–West Corner Rule
2. Row Minima Method
3. Column Minima Method
4. Lowest Cost Entry Method (Matrix Minima Method)
5. Vogel's Approximation Method (Unit Cost Penalty method).

We illustrate Phase 1 for each of the above methods by taking an example. We take an example and solve it by all the five methods just to make a clear comparison. For the benefit of the students, the steps involved in solving are explained along with the problem only. Students can formulate their own algorithms, if necessary.

(*Note* Above all, while solving a transportation problem, the figures given in the problem represent the unit cost of transportation from that origin to that destination, respectively. These values are written down at the right corner of the cell. After the allocations are made, the respective allocated values are to be multiplied with the unit costs of transportation to arrive at the total cost figure.)

**Illustration 4.3**

Solve the following transportation problem (TP) and obtain the transportation cost:

| | | Destination (To) | | | | Availability |
|---|---|---|---|---|---|---|
| | | $D_1$ | $D_2$ | $D_3$ | $D_4$ | ↓ |
| | $S_1$ | 6 | 4 | 1 | 5 | 14 |
| Source (From) | $S_2$ | 8 | 9 | 2 | 7 | 16 |
| | $S_3$ | 4 | 3 | 6 | 2 | 5 |
| Requirement | → | 6 | 10 | 15 | 4 | |

**Solution**

The first step before solving any transportation problem is to check whether $\Sigma A_i = \Sigma B_j$.

In this case, we have $\Sigma A_i = 35$, $\Sigma B_j = 35$ $\therefore$ $\Sigma A_i = \Sigma B_j = 35$.

The figures 6, 4, 1, 5...etc., are the costs associated with the transportation of 1 unit from the respective source to the respective destination. For example, 6 is the cost of transportation/unit from $S_1$ to $D_1$ and so on.

(a) North-West Corner Rule

**Table 4.1** Solution as per North-West Corner Rule

| Source (From) \ Destination (To) | $D_1$ | $D_2$ | $D_3$ | $D_4$ | Availability ↓ |
|---|---|---|---|---|---|
| $S_1$ | ⑥ 6 | ⑧ 4 | 1 | 5 | ~~14~~ ~~8~~ 0 |
| $S_2$ | 8 | ② 9 | ④ 2 | 7 | ~~16~~ ~~14~~ 0 |
| $S_3$ | 4 | 3 | ① 6 | ④ 2 | ~~5~~ ~~4~~ 0 |
| Requirement → | ~~6~~ 0 | ~~10~~ ~~2~~ 0 | ~~15~~ ~~1~~ 0 | ~~4~~ 0 | |

## North-West Corner Rule

In the North-West Corner Rule (NWCR) as the name suggests, the first allocation is made to North-West Corner cell, i.e., the left topmost cell $X_{11}$. While allocating the number of units for that particular cell, one has to look at the demand and supply values, and the least of the two is to be allocated in that cell. Here, the demand is 6 and the availability is 14. The least of the two is 6, so we allot 6 units to cell $X_{11}$ and reduce 6 units in the row availability value, i.e., 14. Now, we are left with 8 units to be allocated in that row only.

After 6 units are allocated, we see there is availability of another 8 units in the same row. (*Note* One has to exhaust a row or a column in order to proceed to the next row or column. Here we have exhausted a column, but the row is not exhausted. So we have to proceed in the same row, i.e., to cell $X_{12}$ to give the allocation. If the row had been exhausted with excess in the column, then we had to give allocation to $X_{21}$). To give allocation to $X_{12}$, we again compare the requirements and the availability and choose the minimum. Here, the requirement is 10 and minimum of the availability is 8. Hence, the minimum is 8. So, we allot 8 units to $X_{12}$.

Now, we have an excess in the second column, an excess of 2 units is available in the second column. So we allot 2 units to cell $X_{22}$. By doing this, we have exhausted the first row and the second column. Again, we have 14 units remaining in the second row. This has to be allocated in the same row, i.e., in cell $X_{23}$. So, the total availability will be exhausted completely.

Look at the requirements in the third column. We have 1 unit left out to be allocated in the third column only. We allocate this in cell $X_{33}$. The last allotment is to be made with the available units in the third row. There are 4 units available in the third row. These are allocated to cell $X_{34}$.

Now, all the requirements have been taken care of with the availabilities. The total cost of transportation can be calculated as shown.

The total cost: $(6 \times 6) + (8 \times 4) + (2 \times 9) + (14 \times 2) + (1 \times 6) + (4 \times 2) =$ Rs. 128.

(b) Row Minima Method

**Table 4.2** Solution as per Row Minima Method

| Source (From) | $D_1$ | $D_2$ | $D_3$ | $D_4$ | Availability ↓ |
|---|---|---|---|---|---|
| $S_1$ | 6 | 4 | (14) 1 | 5 | ~~14~~ 0 |
| $S_2$ | (6) 8 | (5) 9 | (1) 2 | (4) 7 | ~~16~~ ~~15~~ ~~11~~ ~~5~~ 0 |
| $S_3$ | 4 | (5) 3 | 6 | 2 | ~~5~~ 0 |
| Requirement → | ~~6~~ 0 | ~~10~~ ~~5~~ 0 | ~~15~~ ~~1~~ 0 | ~~4~~ 0 | |

Destination (To)

This method is another way of finding the transportation schedule. In this method, we just concentrate on the minimum transportation costs in each row, whereas in the NWCR, we started from the North-West Corner cell.

Take a look at the first row. The minimum transportation cost is 1 (cell $X_{13}$). The minimum of the availability and the requirement values is allocated to this cell, i.e., 14 units are allocated to cell $X_{13}$. Hence, the first row is exhausted. We move to the second row.

The minimum transportation cost is 2 (cell $X_{23}$). We have 1 unit remaining to be allocated in the third column. So, we allocate 1 unit to cell $X_{23.}$ The next least cost is 7 (cell $X_{24}$). We have 4 units in the fourth column. This can be allocated to cell $X_{24}$. Now, we have to observe that 5 units available in the second column are allocated in the second row. We can allocate the remaining 11 units in the same row. We move to the cell with the next least cost, i.e., 8 (cell $X_{21}$). We allocate the minimum of {6, 11}, i.e., 6. This leaves another 5 units to be allocated in the same row. This is allocated to the only remaining cell $X_{22}$.

The only column that is not exhausted is the second column. The only cell to which the allocation can be made is to cell $X_{32}$. We allocate the remaining 5 units to this cell. So, the availabilities have met the requirements. Now, we calculate the total cost as shown.

By the Row Minima Method, the total cost of transportation is $= (14 \times 1) + (6 \times 8) + (5 \times 9) + (1 \times 2) + (4 \times 7) + (5 \times 3) =$ Rs. 152.

(*Note* In the Row Minima Method, we have to exhaust one row before proceeding to the next row. We have to start from the first row only and then proceed to successive rows).

(c) Column Minima Method

**Table 4.3** Solution as per Column Minima Method

| | | Destination (To) $D_1$ | $D_2$ | $D_3$ | $D_4$ | Availability ↓ |
|---|---|---|---|---|---|---|
| Source (From) | $S_1$ | ① 6 | ⑩ 4 | ③ 1 | 5 | ~~14~~ ~~13~~ ~~3~~ 0 |
| | $S_2$ | 8 | 9 | ⑫ 2 | ④ 7 | ~~16~~ ~~4~~ 0 |
| | $S_3$ | ⑤ 4 | 3 | 6 | 2 | ~~5~~ 0 |
| Requirement → | | ~~6~~ ~~1~~ 0 | ~~10~~ 0 | ~~15~~ ~~12~~ 0 | ~~4~~ 0 | |

In the Row Minima Method, we select the cell with the least transportation cost in the row. Here, we go column wise, i.e., we select the cell with the least cost in the column. (*Note* We have to exhaust one column completely before proceeding to successive columns.)

The least cost cell in the first column is cell $X_{31}$ with value 4. We allocate the minimum of the requirement and availability values to this cell, i.e., 5. We have one unit in excess in the same column. This is allocated to the next least cost cell in the same column, i.e., to cell $X_{11}$. This ensures that the requirements in the first column are taken care of. We move to the second column.

In the second column, we cannot make any allocation to cell $X_{32}$, as the availabilities in the third row are exhausted. So we select the next least cost cell, i.e., cell $X_{12}$ with value 4. We allocate the minimum of {10, 14}, i.e., 10. The requirement in the second column is exhausted. So, we can move to the third row.

The same procedure is repeated here and we allocate 12 units to cell $X_{23}$, and 3 units to cell $X_{13}$, respectively.

Similarly, we allocate the remaining 4 units to cell $X_{24}$, i.e., in the fourth column.

Now all the allocations are made, and we calculate the total transportation cost.

The total transportation cost in this case is $= (1 \times 6) + (10 \times 4) + (3 \times 1) + (12 \times 1) + (4 \times 7) + (5 \times 4) =$ Rs. 121.

(d) Matrix Minima Method (*Least Cost Cell Method*)

In the previous two methods, we selected the least cost elements either in a row or a column. Here, we select the least cost cell in the whole matrix to make the first allocation. The least cost cell is $X_{13}$ with the value 1. The minimum of {14, 15} is allocated to this cell. After this allocation is made, the next allocation is made to the next least cost cell, i.e., cell $X_{23}$ with value 2. There is an availability of 1 unit in this column, so we allocate 1 unit to cell $X_{23}$, thus exhausting the requirements in the third column.

Similarly, we make the next allocation to the next least cost cell, i.e., cell $X_{32}$, minimum of {5, 10}. In the same way, we make all the other allocations and calculate the transportation cost.

**Table 4.4** Solution as per Matrix Minima Method (*Least Cost Cell Method*)

| Source (From) | $D_1$ | $D_2$ | $D_3$ | $D_4$ | Availability ↓ |
|---|---|---|---|---|---|
| $S_1$ | 6 | 4 | ⑭ 1 | 5 | ~~14~~ 0 |
| $S_2$ | ⑥ 8 | ⑤ 9 | ① 2 | ④ 7 | ~~16~~ ~~15~~ ~~11~~ ~~5~~ 0 |
| $S_3$ | 4 | ⑤ 3 | 6 | 2 | ~~5~~ 0 |
| Requirement → | ~~6~~ 0 | ~~10~~ ~~5~~ 0 | ~~15~~ ~~1~~ 0 | ~~4~~ 0 | |

(Destination (To): $D_1$–$D_4$)

The total cost according to this method would be

$$= (14 \times 1) + (6 \times 8) + (5 \times 9) + (1 \times 2) + (4 \times 7) + (5 \times 3) = \text{Rs. } 152.$$

(e) Vogel's Approximation Method

**Table 4.5** Solution as per Vogel's Approximation Method

| Source (From) | $D_1$ | $D_2$ | $D_3$ | $D_4$ | Availability ↓ | | | | | | |
|---|---|---|---|---|---|---|---|---|---|---|---|
| $S_1$ | 6 | ⑩ 4 | 1 | ④ 5 | ~~14~~ 4 0 | | 3 | 3 ← | 4 ← | |
| $S_2$ | ① 8 | 9 | ⑮ 2 | 7 | ~~16~~ ~~1~~ 0 | | 5 ← | 1 | 1 | 1 |
| $S_3$ | ⑤ 4 | 3 | 6 | 2 | ~~5~~ 0 | 1 | 1 | 1 | 1 | 1 |
| Requirement → | ~~6~~ 5 0 | ~~10~~ ~~0~~ | ~~15~~ ~~0~~ | ~~4~~ 0 | | | | | | |
| | 2 | 1 | 1 | 3 | | | | | | |
| | 2 | 1 | | 3 | | | | | | |
| | 2 | | | 3 | | | | | | |
| | 2 ↑ | | | | | | | | | |

(Destination (To): $D_1$–$D_4$)

This is the most feasible method for transportation scheduling and provides the optimal cost of transportation.

The procedure followed to calculate the transportation cost is given below:

Start from the first row. Take the difference of the least cost element and the next least cost element in the respective row. In the first row, the least cost is 1, and the next least cost is 4. The difference is 3, and we enter this on the right as shown. Repeat this procedure for all the rows and columns (difference of least cost and the next least cost in the respective row and column), and these values are entered as shown below.

From the values entered, select the maximum value and mark ↑, an arrow below this value. This means that the respective row or column is selected to make the first allocation. The allocation is made to the least cost cell in that row or column.

Here, the maximum is 5, i.e., the second row is selected for allocation. In this row, the least cost cell is selected, i.e., cell $X_{23}$ with value 2. The minimum of {16, 15}, i.e., 15, is allocated to this cell. By doing this, the requirement in the third column is exhausted. So we draw a straight line below the third column.

Similarly, the differences are calculated again. Now, to take the difference in the second row the allocated cell, i.e., cell $X_{23}$, should be omitted. Now, the least cost is 7 and the next least cost is 8. The difference between these two values must be taken and entered by the side of the row. Again, the highest difference value is selected and the least cost cell in the respective row or column is selected for making the allocation.

The above procedure is repeated till all the allocations are made. The final figure is shown in Table 4.5.

After all the allocations are made, the transportation cost is calculated. The total cost by VAM would be $= (10 \times 4) + (4 \times 5) + (1 \times 8) + (15 \times 2) + (5 \times 4) =$ Rs. 118. Note that the transportation cost by this method is the least among all the other methods. Hence, it is evident that VAM is the most economical approach for a transportation problem.

## 4.5 Moving Toward Optimality

After the initial BFS is obtained to a given transportation problem, the following steps should be followed for reaching the optimum solution:

Step 1: Examination of IBFS for degeneracy. If there is degeneracy, make it non-degenerate (discussed in Sect. 4.6).

Step 2: 1. Determination of net evaluation for empty cells

2. Optimality test of current solution.

Step 3: If Step 2 above indicates that the current solution can be improved, then selection of entering variable.

Step 4: Selection of leaving variable.
Step 5: Repeat steps 1–4 until an optimal solution is obtained.

### 4.5.1 Determination of Net Values (U, V Method)

U, V method is also known as method of multipliers. In this method, the multipliers $U_i$ and $V_j$ are associated with row $i$ and the column $j$ of the transportation table, for each basic variable (cell which has an allocation). The multipliers $U_i$ and $V_j$ must satisfy the equation, $U_i + V_j = C_{ij}$, for each of the basic cells $X_{ij}$.

The initial value of one of the multipliers is assumed to be zero ($U_1 = 0$).

From this, all the $U$, $V$ values are obtained. After all the $U$, $V$ values are calculated, the evaluation of the non-basic variable (cells without allocations or empty cells) is obtained by the formula

$$\hat{C}_{pq} = U_p + V_q - C_{pq}.$$

The entering variable is then selected as the non-basic variable with the most positive $\hat{C}_{pq}$ value.

**Illustration 4.4**
Find the optimum solution in the following transportation problem:

| | | Destination (To) | | | | Availability |
|---|---|---|---|---|---|---|
| | | $D_1$ | $D_2$ | $D_3$ | $D_4$ | ↓ |
| | $S_1$ | 20 | 0 | 40 | 22 | 30 |
| Source (From) | $S_2$ | 24 | 14 | 18 | 40 | 50 |
| | $S_3$ | 0 | 28 | 32 | 36 | 10 |
| Requirement | → | 10 | 30 | 30 | 20 | |

**Solution**
The first step is to check whether $\Sigma A_i = \Sigma B_j$. Here, we have $\Sigma A_i = \Sigma B_j = 90$.

By NWCR, we get the optimal solution as shown in the following table.

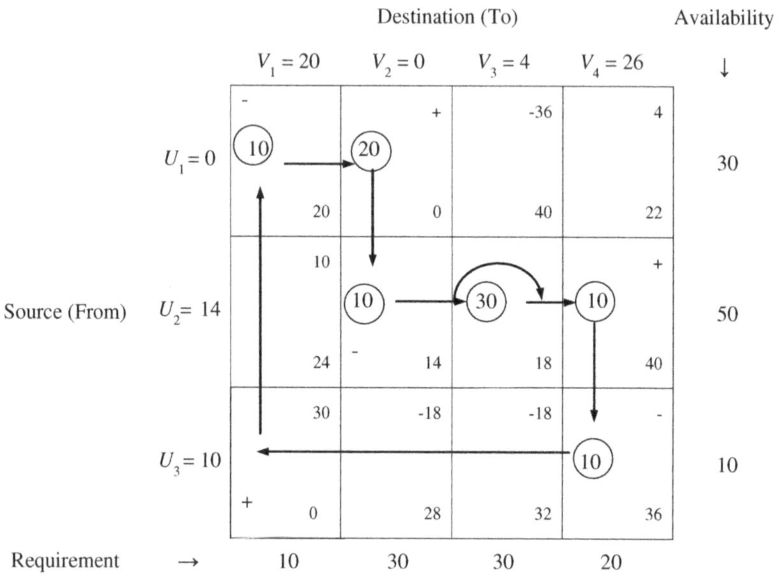


### The calculation of *U*, *V* values

We assume $U_1 = 0$. One thing to remember is, these *U* and *V* vales are calculated against only assigned cells. We have the formula $U_i + V_j = C_{ij}$.

For cell (1, 1) or $X_{11} = U_1 = 0, V_1 = ?, C_{11} = 20$

$$0 + V_1 = 20 \quad \therefore V_1 = 20$$

### Loop Construction

The starting point for the loop will be the cell that has got the highest positive net evaluation, i.e., $X_{31}$. We move upwards in the clockwise direction, touch $X_{11}$, one of the corner points, take a right turn, touch $X_{22}$, take a right turn to $X_{22}$, and take left to $X_{23}$. Now there are two allocated cells, cell $X_{23}$ will lead us nowhere and the loop cannot be completed. Hence, we jump and touch $X_{24}$, take a right to $X_{34}$ and take a right that completes the loop.

The next step is to alternate + and − signs at each corner of the loop, with + starting from the initial corner points, i.e., $X_{31}$.

Select the minimum values among the (−) sign corner points. Here, we have three cells with same values, i.e., 10. In these cases, we can select any one cell, we select $X_{11}$. Shift the value to $X_{31}$ and now $X_{11}$ becomes empty as shown in the next table.

**Calculation of $U$, $V$ values, $U_1 = 0$**

$$\text{Cell } X_{12} = U_1 + V_2 = 0 = 0 + V_2 = 0 \quad \therefore V_2 = 0$$
$$\text{Cell } X_{22} = U_2 + V_2 = 14 = 0 + U_2 = 14 \quad \therefore U_2 = 14$$
$$\text{Cell } X_{23} = U_2 + V_3 = 18 = 14 + V_3 = 18 \quad \therefore V_3 = 4$$
$$\text{Cell } X_{24} = U_2 + V_4 = 40 = 14 + V_4 = 40 \quad \therefore V_4 = 26$$
$$\text{Cell } X_{34} = U_3 + V_4 = 36 = 26 + U_3 = 36 \quad \therefore U_3 = 10$$

We represent the above obtained $U$, $V$ values as shown in the next table.

**Calculation of net evaluations**

Net evaluations are calculated only for empty cell $X_{pq}$. We have the formula $\overline{C_{pq}} = U_p + V_q - C_{pq}$.

$$\text{For Cell } X_{13}, \text{we have } \overline{C_{13}} = 0 + 4 - 40 = -36$$
$$\text{For Cell } X_{14}, \overline{C_{14}} = 0 + 26 - 22 = 4$$
$$\text{For Cell } X_{21}, \overline{C_{21}} = 14 + 20 - 24 = 10$$
$$\text{For Cell } X_{31}, \overline{C_{31}} = 10 + 20 - 0 = 30$$
$$\text{For Cell } X_{32}, \overline{C_{32}} = 10 + 0 - 28 = -18$$
$$\text{For Cell } X_{33}, \overline{C_{33}} = 10 + 4 - 32 = -18.$$

We represent these values on the right-hand top corner of each empty cell. Now, we examine these values. Presence of positive values shows that there is scope for improving the optimal solution.

| | | Destination (To) | | | | Availability |
|---|---|---|---|---|---|---|
| | | $V_1 = -10$ | $V_2 = 0$ | $V_3 = 4$ | $V_4 = 26$ | ↓ |
| | $U_1 = 0$ | -30<br>20 | -<br>(30)<br>0 | -36<br>40 | + 4<br>22 | 30 |
| Source (From) | $U_2 = 14$ | 10<br>24 | (0)<br>+<br>14 | (30)<br>18 | (20)<br>40 | 50 |
| | $U_3 = 10$ | 30<br>0 | -18<br>28 | -18<br>32 | (0)<br>36 | 10 |
| Requirement | → | 10 | 30 | 30 | 20 | |

**Calculation of $U$, $V$ values, $U_1 = 0$**

$$\text{Cell}\,X_{12} = U_1 + V_2 = 0 = 0 + V_2 = 0 \quad \therefore V_2 = 0$$
$$\text{Cell}\,X_{22} = U_2 + V_2 = 14 = 0 + U_2 = 14 \quad \therefore U_2 = 14$$
$$\text{Cell}\,X_{23} = U_2 + V_3 = 18 = 14 + V_3 = 18 \quad \therefore V_3 = 4$$
$$\text{Cell}\,X_{24} = U_2 + V_4 = 40 = 14 + V_4 = 40 \quad \therefore V_4 = 26$$
$$\text{Cell}\,X_{34} = U_3 + V_4 = 36 = 26 + U_3 = 36 \quad \therefore U_3 = 10$$
$$\text{Cell}\,X_{31} = U_3 + V_1 = 0 = 10 + V_1 = 0 \quad \therefore V_3 = -10$$

**Calculation of net evaluations**

$$\text{For Cell}\,X_{11}, \overline{C_{11}} = 0 - 10 - 20 = -30$$
$$\text{For Cell}\,X_{13}, \overline{C_{13}} = 0 + 4 - 40 = -36$$
$$\text{For Cell}\,X_{21}, \overline{C_{21}} = 14 + 10 - 24 = -20$$
$$\text{For Cell}\,X_{32}, \overline{C_{32}} = 10 + 0 - 28 = -18$$
$$\text{For Cell}\,X_{33}, \overline{C_{33}} = 10 + 4 - 32 = -18$$
$$\text{For Cell}\,X_{14}, \overline{C_{14}} = 0 + 26 - 22 = 4.$$

Now, we see that there is one cell with positive value. We take this as initial corner point for loop construction and construct the loop as shown in the previous table. The next iteration would be:

Destination (To)

| Source (From) | $V_1 = -14$ | $V_2 = 0$ | $V_3 = 4$ | $V_4 = 22$ | Availability ↓ |
|---|---|---|---|---|---|
| $U_1 = 0$ | -30 / 20 | (10) / 0 | -36 / 40 | 4 / (20) / 22 | 30 |
| $U_2 = 14$ | 10 / 24 | (0) / 14 | (30) / 18 | 40 | 50 |
| $U_3 = 14$ | 30 / (10) / 0 | -18 / 28 | -18 / 32 | (0) / 36 | 10 |
| Requirement → | 10 | 30 | 30 | 20 | |

We determine the $U$, $V$ values and the net evaluations for empty cells as shown in previous Table. We see that all net evaluations are negative. This indicates that the obtained solution is optimal.

Transportation cost $= (20 \times 22) + (30 \times 18) + 10 =$ Rs. 980.

**Illustration 4.5**

A company has factories $F_1$, $F_2$, and $F_3$ which supply warehouses at $W_1$, $W_2$, and $W_3$. Weekly factory capacities are 200, 160, and 90 units, respectively. Weekly warehouse requirements are 180, 120, and 150 units, respectively. Unit shipping costs (in rupees) are as follows in below table.

| | Warehouse | | | |
|---|---|---|---|---|
| | $W_1$ | $W_2$ | $W_3$ | Supply |
| $F_1$ | 16 | 20 | 12 | 200 |
| $F_2$ | 14 | 8 | 18 | 160 |
| $F_3$ | 26 | 24 | 16 | 90 |
| Demand | 180 | 120 | 150 | 350 |

Determine the optimum distribution for this company that minimizes the shipping costs.

**Solution**

By VAM the IBFS is obtained as shown in the below table.

| | | Destination (To) | | | Availability ↓ |
|---|---|---|---|---|---|
| | | $V_1 = 16$ | $V_2 = 10$ | $V_3 = 12$ | |
| Source (From) | $U_1 = 0$ | (140) 16 | [-10] 20 | (60) 12 | 200 |
| | $U_2 = -2$ | (40) 14 | (120) 8 | [-8] 18 | 160 |
| | $U_3 = 4$ | [-6] 26 | [-10] 24 | (90) 16 | 9 |
| Requirement → | | 180 | 120 | 150 | |

As we can see from above table, the net evaluation of all empty cells is either negative or zero. So the IBFS is the optimal solution.

$$\text{The transportation cost} = (140 \times 16) + (60 \times 12) + (40 \times 14) + (120 \times 8) + (90 \times 16)$$
$$= \text{Rs. } 5{,}920.$$

## 4.6 Degeneracy in Transportation Problems

We have discussed the solution methodology for non-degenerate BFS with exactly $m + n - 1$ strictly positive allocations in "independent" positions. Sometimes, it is not possible to get such an initial BFS at the start. Degeneracy, in transportation problems, hence occurs whenever the number of occupied cells is less than $m + n-1$. Degeneracy can occur when

(a) BFS is degenerate from the initial stages
(b) It may become degenerate at an intermediate stage.

### 4.6.1 Checking the Initial BFS for Degeneracy

If the number of allocations is less than $m + n - 1$, then the transportation problem is said to be degenerate, where $m$ is the number of rows and $n$ the number of columns. Degeneracy may set in either in the initial stage of the solution itself or on successive iterations, i.e., in the process of converting the IBFS to the optimal basic feasible solution (OBFS).

The following procedure is adopted to resolve degeneracy in transportation problems:

1. The first step is to identify the number by which the initial allocations fall short of the required number of allocations, i.e., $m + n - 1$.
2. Next, locate the number of unoccupied cells to make the number of allocations $m + n - 1$. These cells must be selected in such a way that they themselves do not result in formation of a closed loop.
3. A small quantity $\in$ is allocated to these independent positions or the unoccupied cells. These allocated cells play the role of a basic cell till optimality is reached. However, $\in$ value is taken to be zero for calculating the transportation costs.

### 4.6.2 Resolving Degeneracy at the Initial Stage

To resolve degeneracy at the initial stage, an extremely small amount of goods, €, is allocated to one or more empty cells so that the number of occupied cells become $m + n - 1$. We illustrate this by an example.

**Illustration 4.6**

A state-government-owned firm has 4 plants: *I*, *N*. *F*, and *Y* and 3 warehouses: *W*, *I*, and *P*. The demand figures are 240, 160, and 200 and supply figures are 100, 160, 140, and 200, respectively. Find the minimum distribution cost, given the unit transportation costs, by NWCR

| | | Destination (To) | | | Availability ↓ |
|---|---|---|---|---|---|
| | | *W* | *I* | *P* | |
| | *I* | 2 | 4 | 6 | 100 |
| Source (From) | *N* | 6 | 4 | 2 | 160 |
| | *F* | 8 | 10 | 12 | 140 |
| | *Y* | 6 | 2 | 4 | 200 |
| Requirement | → | 240 | 160 | 200 | |

**Solution**

By NWCR, the IBFS is attained as shown in the table below.

| | | Destination (To) | | | Availability ↓ |
|---|---|---|---|---|---|
| | | *W* | *I* | *P* | |
| | *I* | (100) 2 | 4 | 6 | 100 |
| Source (From) | *N* | (140) 6 | (20) 4 | 2 | 160 |
| | *F* | 8 | (140) 10 | 12 | 140 |
| | *Y* | (ε) 6 | 2 | (200) 4 | 200 |
| Requirement | → | 240 | 160 | 200 | |

In above table, there are 4 rows ($m$) and 3 columns ($n$). The number of allocations is 5. We have

$$m + n - 1 = 4 + 3 - 1 = 6 \neq 5$$

Therefore, it is a degenerate transportation problem.

(*Note* Once the IBFS is obtained, one must check for degeneracy. Solving without checking for degeneracy will result in no solution).

In order to resolve degeneracy, a small quantity $\in$ is added as shown in the previous table. The allocation is made such that this cell $X_{41}$ does not result in formation of closed loop within the existing allocations.

Now, we calculate $U$, $V$ values and net evaluation and proceed in the usual manner toward OBFS.

| Source (From) | | Destination (To) $V_1 = 2$ | $V_2 = 0$ | $V_3 = 0$ | Availability ↓ |
|---|---|---|---|---|---|
| | $U_1 = 0$ | (100) 2 | [-4] 4 | [-6] 6 | 100 |
| | $U_2 = 4$ | (140) 6 | (20) 4 | [+2] 2 | 160 |
| | $U_3 = 10$ | [4] 8 | (140) 10 | [-2] 12 | 140 |
| | $U_4 = 4$ | ($\in$) 6 | [2] 2 | (200) 4 | 200 |
| Requirement | → | 240 | 160 | 200 | |

| Source (From) | | Destination (To) $V_1 = 2$ | $V_2 = 4$ | $V_3 = 0$ | Availability ↓ |
|---|---|---|---|---|---|
| | $U_1 = 0$ | (100) 2 | [0] 4 | [-6] 6 | 100 |
| | $U_2 = 0$ | [-2] 6 | (160) 4 | [-2] 2 | 160 |
| | $U_3 = 6$ | (140) 8 | (0) 10 | [-6] 12 | 140 |
| | $U_4 = 4$ | ($\in$) 6 | [6] 2 | (100) 4 | 200 |
| Requirement | → | 240 | 160 | 200 | |

As seen, we reach optimality in bottom table, as all net evaluations are negative. The transportation cost = Rs. 2,120.

Destination (To) / Availability

| Source (From) | $V_1 = 2$ | $V_2 = 4$ | $V_3 = 6$ | Availability ↓ |
|---|---|---|---|---|
| $U_1 = 0$ | (100) 2 | [0] 4 | [0] 6 | 100 |
| $U_2 = 0$ | [-4] 6 | (160) 4 | [4] 2 | 160 |
| $U_3 = 6$ | (140) 8 | (0) 10 | [0] 12 | 140 |
| $U_4 = -2$ | [-6] 6 | (∈) 2 | (200) 4 | 200 |
| Requirement → | 240 | 160 | 200 | |

Destination (To) / Availability

| Source (From) | $V_1 = 2$ | $V_2 = 4$ | $V_3 = 6$ | Availability ↓ |
|---|---|---|---|---|
| $U_1 = 0$ | (100) 2 | [0] 4 | [0] 16 | 100 |
| $U_2 = -4$ | [-2] 6 | [-4] 4 | (160) 2 | 160 |
| $U_3 = 6$ | (140) 8 | (0) 10 | [0] 12 | 140 |
| $U_4 = -2$ | [-6] 6 | (100) 2 | (40) 4 | 200 |
| Requirement → | 240 | 160 | 200 | |

### *4.6.3 Resolving Degeneracy During Solution Stages*

A transportation problem may also become degenerate at solution stages. It may happen due to most favorable quantity being allocated to empty cell with largest negative cell evaluations resulting in simultaneous variation of the two or more occupied cells. To resolve this degeneracy, a small quantity, $\delta$, is allocated to one or more of recently vacated cells so that the number of occupied cells is $m + n - 1$ in the new solution. This is illustrated with examples.

**Illustration 4.7**

Given the cost requirement table, what would be your optimal transportation plan? Obtain the IBFS by VAM and give the minimum distribution cost.

| Source (From) | Designation (To) | $W_1$ | $W_2$ | $W_3$ | $W_4$ | $W_5$ | $W_6$ | Availability ↓ |
|---|---|---|---|---|---|---|---|---|
| | $F_1$ | 27 | 36 | 27 | 18 | 27 | 30 | 15 |
| | $F_2$ | 21 | 9 | 21 | 21 | 15 | 15 | 18 |
| | $F_3$ | 18 | 15 | 27 | 36 | 27 | 33 | 6 |
| | $F_4$ | 18 | 24 | 33 | 6 | 6 | 30 | 27 |
| Requirement | → | 12 | 12 | 18 | 6 | 12 | | |

**Solution**

Below table gives the IBFS by VAM.

| Source (From) | Destination: $V_1 = 18$ | $V_2 = 15$ | $V_3 = 27$ | $V_4 = 6$ | $V_5 = 6$ | $V_6 = 21$ | Availability ↓ |
|---|---|---|---|---|---|---|---|
| $U_1 = 0$ | (-9) 27 | (-21) 36 | [15] 27 | (-12) 18 | (-21) 27 | (-9) 30 | 15 |
| $U_2 = -6$ | (-9) 21 | [12] 9 | [∈] 21 | (-21) 21 | (-15) 15 | [6] 15 | 18 |
| $U_3 = 0$ | [3] 18 | (0) 15 | [3] 27 | (-30) 36 | (-21) 27 | (-12) 33 | 6 |
| $U_4 = 0$ | [9] 18 | (-9) 24 | (-6) 33 | [6] 6 | [12] 6 | (-9) 30 | 27 |
| Requirement | 12 | 12 | 18 | 6 | 12 | 6 | |

As we can see from above table, $m + n - 1 = 6 + 4 - 1 = 9 \neq 8$. So the IBFS is degenerated. We introduce a small quantity in cell $X_{23}$ so that no closed-loop formation is possible.

Since the net evaluation of all the empty cells is negative, the IBFS itself is the optimum transportation plan. The minimum distribution cost is

$$= (15 \times 27) + (12 \times 9) + (3 \times 8) + (3 \times 27) + (9 \times 8) + (6 \times 6) + (12 \times 6)$$
$$= \text{Rs. } 1{,}608$$

(*Note* There is no need to show $\in$ in calculating the cost, as $\in = 0$).

## 4.7 Maximization in Transportation Problems

Sometimes, we may come across situations wherein the transportation problems are expressed in maximization form. In order to solve these kinds of problems, the given problem must first be converted to minimization form before applying the TP technique. For conversion from maximization to minimization, subtract all the elements in the transportation matrix from the highest element in the matrix and rewrite the minimization matrix of the transportation problem. We illustrate this with an example.

**Illustration 4.8**
A company owns three plants $P$, $Q$, and $R$ and three outlets $X$, $Y$, and $Z$, where manufacturing and selling activities take place, respectively. The ex-factory profit matrix and unit costs of transportation are given in below tables, respectively. What is the optimal distribution schedule?

| | | Destination (To) | | |
|---|---|---|---|---|
| | | $X$ | $Y$ | $Z$ |
| Source (From) | $P$ | 500 | 1000 | 1500 |
| | $Q$ | 1500 | 1000 | 500 |
| | $R$ | 2000 | 2500 | 3000 |

| | | Destination (To) | | | Availability ↓ |
|---|---|---|---|---|---|
| | | $X$ | $Y$ | $Z$ | |
| Source (From) | $P$ | 50 | 100 | 150 | 2000 |
| | $Q$ | 200 | 150 | 100 | 500 |
| | $R$ | 250 | 300 | 350 | 500 |
| Requirement | → | 500 | 1000 | 1500 | |

**Solution**

Given the ex-factory profit matrix and unit costs of transportation, we can calculate the net profit matrix as

$$\text{Net profit} = \text{Ex-factory profit} - \text{Unit cost of transportation}$$

Destination (To)

| | | X | Y | Z | Availability ↓ |
|---|---|---|---|---|---|
| Source (From) | P | 500 – 50 = 450 | 1000 – 100 = 900 | 1500 – 150 = 1350 | 2000 |
| | Q | 1500 – 200 = 1300 | 1000 – 150 = 850 | 500 – 100 = 400 | 500 |
| | R | 2000 – 250 = 1750 | 2500 – 300 = 2200 | 3000 – 350 = 2650 | 500 |
| Requirement → | | 500 | 1000 | 1500 | |

The optimum distribution schedule cannot be calculated from the net profit matrix. It has to be converted to a loss matrix. We do this by subtracting all the cost elements from the highest cost element, i.e., 2,650 as shown in the below table.

Destination (To)

| | | X | Y | Z | Availability ↓ |
|---|---|---|---|---|---|
| Source (From) | P | 2650 – 450 = 2200 | 2650 – 900 = 1750 | 2650 – 1350 = 1300 | 2000 |
| | Q | 2650 – 1300 = 1350 | 2650 – 850 = 1800 | 2650 – 400 = 2250 | 500 |
| | R | 2650 – 1750 = 900 | 2650 – 2200 = 450 | 2650 – 2650 = 0 | 500 |
| Requirement → | | 500 | 1000 | 1500 | |

Now, we obtain as IBFS through VAM as shown in the table below.

Destination (To)

| | | $V_1 = 2200$ | $V_2 = 1750$ | $V_3 = 1300$ | Availability ↓ |
|---|---|---|---|---|---|
| Source (From) | P | [∈] 2200 | [500] 1750 | [1500] 1300 | 2000 |
| | Q | [500] 1350 | -900 / 1800 | -1800 / 2250 | 500 |
| | R | 0 / 900 | [450] 450 | 0 / 0 | 500 |
| Requirement → | | 180 | 120 | 150 | |

We examine $m + n - 1 = 3 + 3 - 1 = 6 \neq 5$. This is a case of degeneracy. Hence, we introduce $\in$ in cell $X_{11}$ to resolve degeneracy.

As the net evaluation of the empty cells is either zero or negative, the optimality condition is satisfied. The previous table indicates optimum distribution schedule. The optimum distribution cost is

$$= (500 \times 1,750) + (500 \times 1,350) + (450 \times 450) + (1,500 \times 1,300)$$
$$= \text{Rs. } 37,02,500$$

## 4.8 Unbalanced Transportation Problem

If in a transportation problem, the sum of all variable quantities is not equal to the sum of the requirements, i.e., it is referred to as unbalanced transportation problem.

$$\sum_{i=1}^{M} A_i \neq \sum_{j=1}^{N} B_j$$

An unbalanced transportation problem may occur due to:

(a) Excess of availability
(b) Shortage of availability.

### *4.8.1 Excess of Availability ($\Sigma A_i \geq \Sigma B_j$)*

In this case, we introduce a dummy destination column in the transportation table. The unit costs to this dummy variable are set equal to zero. The requirement at this dummy destination is assumed to be equal to the difference $\Sigma A_i - \Sigma B_j$.

**Illustration 4.9**
Solve the following TP given the transportation cost, and find the optimal shipping schedule and cost.

| | | Destination (To) | | | | | Availability |
|---|---|---|---|---|---|---|---|
| | | $D_1$ | $D_2$ | $D_3$ | $D_4$ | $D_5$ | ↓ |
| Source (From) | $O_1$ | 8 | 4 | 6 | 4 | 12 | 16 |
| | $O_2$ | 10 | 8 | 10 | 4 | 12 | 24 |
| | $O_3$ | 12 | 10 | 8 | 14 | 6 | 22 |
| Requirement | → | 8 | 8 | 12 | 16 | 16 | |

**Solution**

First, we check if $\Sigma A_i = \Sigma B_j$, $\Sigma A_i = 60$, $\Sigma B_j = 68$.

Therefore, $\Sigma A_i \neq \Sigma B_j$ or $\Sigma A_i < \Sigma B_j$ or $\Sigma B_j - \Sigma A_i = 8$.

So, this is an unbalanced transportation problem where the requirement is less than availability. In this case, we modify the problem by adding a dummy destination with 0 as the unit cost of transportation.

| | | Destination (To) | | | | | | Availability |
|---|---|---|---|---|---|---|---|---|
| | | $D_1$ | $D_2$ | $D_3$ | $D_4$ | $D_5$ | $D_6$ | ↓ |
| Source (From) | $O_1$ | 8 | 4 | 6 | 4 | 12 | 0 | 16 |
| | $O_2$ | 10 | 8 | 10 | 4 | 12 | 0 | 24 |
| | $O_3$ | 12 | 10 | 8 | 14 | 6 | 0 | 28 |
| Requirement | → | 8 | 8 | 12 | 16 | 16 | 8 | |

Now, $\Sigma A_i = \Sigma B_j = 68$.

We obtain the IBFS by VAM as shown in the table below.

| Source (From) | Destination $V_1 = 8$ | $V_2 = 4$ | $V_3 = 4$ | $V_4 = 4$ | $V_5 = 2$ | $V_6 = -4$ | Availability ↓ |
|---|---|---|---|---|---|---|---|
| $U_1 = 0$ | [∈] 8 | [8] 4 | -2 / 6 | [8] 4 | -10 / 12 | -4 / 0 | 16 |
| $U_2 = -6$ | -2 / 10 | -4 / 8 | -6 / 10 | [8] 4 | [16] 2 | -4 / 0 | 24 |
| $U_3 = 0$ | [8] 12 | -2 / 10 | [12] 8 | -6 / 14 | 0 / 6 | [8] 0 | 28 |
| Requirement | 12 | 12 | 18 | 6 | 12 | 6 | |

Next, we check $m + n - 1 = 6 + 3 - 1 = 8$. But $8 \neq 7$ (number of allocations).

So degeneracy exists here, and we allocate a small quantity $\in$ to cell $X_{11}$ to resolve degeneracy and calculate $U$, $V$ values and net evaluations.

Since, the net evaluations of all empty cells in the last table of the Sect. 4.8.2 are either zero or negative, the above table gives the optimal shipping schedule.

The optimal cost of shipping is

$$= (8 \times 4) + (8 \times 4) + (8 \times 4) + (16 \times 2) + (8 \times 12) + (12 \times 8) = \text{Rs. } 320.$$

### 4.8.2 Shortage of Availability ($\Sigma A_i \leq \Sigma B_j$)

In this case, we introduce a dummy source in the transportation table. The costs of transportation from this dummy source to any destination are all set equal to zero. The availability at this source is assumed to be equal to the difference ($\Sigma B_j - \Sigma A_i$).

**Illustration 4.10**
Consider the following transportation matrix:

| Source (From) | | Destination (To) 1 | 2 | 3 | Availability ↓ |
|---|---|---|---|---|---|
| | 1 | 15 | 3 | 21 | 30 |
| | 2 | 18 | 12 | 18 | 240 |
| | 3 | 9 | 6 | 15 | 45 |
| Requirement → | | 225 | 60 | 150 | |

**Solution**
In this case, the demand exceeds supply by 120 units. So we add a dummy row or destination of 120 units availability with zero cost value.

| Source (From) | | Destination (To) 1 | 2 | 3 | Availability ↓ |
|---|---|---|---|---|---|
| | 1 | 15 | 3 | 21 | 30 |
| | 2 | 18 | 12 | 18 | 240 |
| | 3 | 9 | 6 | 15 | 45 |
| | 4 | 0 | 0 | 0 | 120 |
| Requirement → | | 225 | 60 | 150 | |

Now we solve by NWCR.

| Source (From) | | Destination (To) 1 | 2 | 3 | Availability ↓ |
|---|---|---|---|---|---|
| | 1 | [30] 15 | 3 | 21 | ~~30~~ 0 |
| Source (From) | 2 | [195] 18 | [45] 12 | 18 | ~~240~~ ~~45~~ 0 |
| | 3 | 9 | [15] 6 | [30] 15 | ~~45~~ ~~30~~ 0 |
| | 4 | 0 | 0 | [120] 0 | ~~120~~ 0 |
| Requirement | → | ~~225~~ ~~195~~ 0 | ~~60~~ ~~15~~ 0 | ~~150~~ 0 | |

The total transportation list is

$$(30 \times 15) + (195 \times 18) + (45 \times 12) + (15 \times 6) + (30 \times 15) = \text{Rs. } 5{,}040.$$

## 4.9 Transshipment Problem

A transportation problem in which the available commodity frequently moves from one source to another source or destination before reaching its actual destination is referred to as transshipment problem.

A transshipment problem can be represented as in Table 4.6.

**Table 4.6** Transshipment Problem

| | $S_1$ | $S_2$ | … | $S_j$ | $D_1$ | $D_2$ | $D_j$ | Supply ↓ |
|---|---|---|---|---|---|---|---|---|
| $S_1$ | | | | | | | | $Su_1$ |
| $S_2$ | | | | | | | | $Su_2$ |
| \| | | | | | | | | . |
| $S_j$ | | | | | | | | . |
| $D_1$ | | | | | | | | . |
| $D_2$ | | | | | | | | . |
| $D_j$ | | | | | | | | $Su_j$ |
| Demand → | $De_1$ | $De_2$ | $De_3$ | … | … | … | $De_i$ | |

| | | Designation | | | | | | Availability/Supply |
|---|---|---|---|---|---|---|---|---|
| | | $S_1$ | $S_2$ | $S_3$ | $S_4$ | $D_2$ | $D_3$ | ↓ |
| Source (From) | $S_1$ | 0 | 8 | 40 | 10 | 50 | 24 | 200 |
| | $S_2$ | 20 | 0 | 12 | 20 | 10 | 40 | 400 |
| | $S_3$ | 30 | 40 | 0 | 16 | 90 | 14 | 300 |
| | $S_4$ | 40 | 50 | 20 | 0 | 60 | 12 | 700 |
| | $D_2$ | 40 | 36 | 120 | 30 | 0 | 20 | |
| | $D_3$ | 20 | 50 | 60 | 46 | 8 | 0 | |
| Requirement/Demand | → | | | | | 700 | 900 | |

**Illustration 4.11**
Solve the following transshipment problem given the unit cost of transportation between various sources and destinations.

**Solution**
Here, the first requirement is to check whether $\sum_{i=1}^{n} a_i = \sum_{j=1}^{n} b_j$. We have $\sum_{i=1}^{4} a_i = \sum_{j=1}^{2} b_j = 1,600$. Now, we have to fill up the demand values for supplies. To do this, we fill up the value of $\sum_{i=1}^{4} a_i = \sum_{j=1}^{2} b_j = 1,600$ for the sources and add this availability as shown in below table.

| | | Designation (To) | | | | | | | Availability |
|---|---|---|---|---|---|---|---|---|---|
| | | $S_1$ | $S_2$ | $S_3$ | $S_4$ | $D_2$ | $D_3$ | | ↓ |
| Source (From) | $S_1$ | 0 | 8 | 40 | 10 | 50 | 24 | → | 1600 + 200 = 1800 |
| | $S_2$ | 20 | 0 | 12 | 20 | 10 | 40 | → | 1600 + 400 = 2000 |
| | $S_3$ | 30 | 40 | 0 | 16 | 90 | 14 | → | 1600 + 300 = 1900 |
| | $S_4$ | 40 | 50 | 20 | 0 | 60 | 12 | → | 1600 + 700 = 2300 |
| | $D_2$ | 40 | 36 | 120 | 30 | 0 | 20 | → | 1600 |
| | $D_3$ | 20 | 50 | 60 | 46 | 8 | 0 | → | 1600 |
| Requirement | → | ↓ | ↓ | ↓ | ↓ | ↓ | ↓ | | |
| | | 1600 | 1600 | 1600 | 1600 | 700+ 1600 = 2300 | 900+ 1600 = 2500 | | |

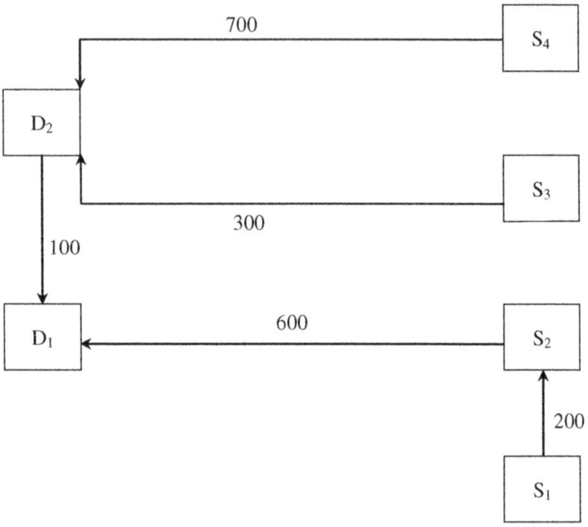


**Fig. 4.1** The optimal shipping pattern

By VAM, we arrive at the optimal allocation as shown in below table.

| | | Designation | | | | | | Availability ↓ |
|---|---|---|---|---|---|---|---|---|
| | | $S_1$ | $S_2$ | $S_3$ | $S_4$ | $D_2$ | $D_3$ | |
| Source (From) | $S_1$ | [1600] 0 | [200] 8 | 40 | 10 | 50 | 24 | 1800 |
| | $S_2$ | 20 | [1400] 0 | 12 | 20 | [600] 10 | 40 | 2000 |
| | $S_3$ | 20 | 40 | [1600] 0 | 16 | 90 | [300] 14 | 1900 |
| | $S_4$ | 40 | 50 | 20 | [1600] 0 | 60 | [700] 12 | 2300 |
| | $D_2$ | 40 | 36 | 120 | 30 | [1600] 0 | 20 | 1600 |
| | $D_3$ | 20 | 50 | 60 | 46 | [100] 8 | [1500] 0 | 1600 |
| Requirement/Demand | → | 1600 | 1600 | 1600 | 1600 | 2300 | 2500 | |

$$\begin{aligned}\text{The transportation cost is} &= (1,600 \times 0) + (200 \times 8) + (1,400 \times 0) + (1,600 \times 0) + (600 \times 10) + (300 \times 14) \\ &\quad + (1,600 \times 0) + (700 \times 12) + (1,600 \times 0) + (100 \times 8) + (1,500 \times 0) \\ &= \text{Rs. } 10{,}500\end{aligned}$$

The optimal shipping pattern is shown in Fig. 4.1.

## 4.10 Generalized Transportation Problem

The contribution of the scientific community to this world has been far beyond one's expectation. Scientific research has enabled development of algorithms for generalization of the transportation problem; however, the complexity of the same does not enable us to include them all in this book. In a generalized transportation problem, the easy way of solving a TP cannot be applied. The algorithms should be used in this case. However, a brief idea is given for the benefit of the student. For a generalized transportation problem, the constraints can be expressed as

$$\sum_{i=1}^{n} A_{ij}X_{ij} = A_i \ldots\ldots \quad \forall i = 1, 2, \ldots m$$

$$\sum_{i=1}^{m} B_{ij}X_{ij} = A_i \ldots\ldots \quad \forall j = 1, 2, \ldots n$$

$$\sum_{i=1}^{m} A_i = \sum_{j=1}^{n} B_i, \quad \text{and} \quad X_{ij} \geq 0$$

## 4.11 Assignment Problem

Assignment problem is a special kind of transportation problem where each source should have the capacity to fulfill the destinations. If any destination is expressed in terms of a shop floor situation, an operator should be able to perform any job regardless of his skills, though the cost incurred may be more when the job and skill set do not match.

Suppose there are $n$ jobs to be performed, and there are $n$ persons available for doing them. Assuming that each person can do each job at a time, though with variation in efficiency, in an assignment problem, the objective is to find an assignment that minimizes the total cost of performing all jobs; we can express this problem as in Table 4.7.

**Table 4.7** Assignment Problem

| | 1 | 2 | ... | $j$ | ... | $n$ |
|---|---|---|---|---|---|---|
| 1 | $C_{11}$ | $C_{12}$ | ... | $C_{ij}$ | ... | $C_{in}$ |
| 2 | $C_{21}$ | $C_{22}$ | ... | $C_{2j}$ | ... | $C_{2n}$ |
| 3 | $C_{i1}$ | $C_{i2}$ | ... | $C_{ij}$ | ... | $C_{in}$ |
| $n$ | $C_{n1}$ | $C_{n2}$ | ... | $C_{nj}$ | ... | $C_{nn}$ |

Mathematically, we can state the assignment problem as:

Minimize the total cost:

$$Z = \sum_{i=1}^{n} \sum_{j=1}^{n} C_{ij} X_{ij} \ldots\ldots \quad \forall i, j = 1, 2, \ldots n$$

subject to

$$X_{ij} = \begin{cases} 1 \text{ if the } i\text{th person is assigned to the } j\text{th job.} \\ 0 \text{ if not} \end{cases}$$

$$\sum_{j=1}^{n} X_{ij} = 1 \text{ (one job is done by the } i\text{th person), and}$$

$$\sum_{i=1}^{n} X_{ij} = 1 \text{ (one person only is to be assigned to the } j\text{th job)}$$

(*Note* $X_{ij}$—jth job is assigned to the *i*th person).

### 4.11.1 Types of Assignment Problems

Assignment problems can be of two types.

(a) Balanced [where rows (jobs) equal columns (persons)] and
(b) Unbalanced [where rows (jobs) do not equal columns (persons)].

In the case of an unbalanced transportation problem, we balance it adding row(s)/col(s) as in an unbalanced transportation problem. The values for the dummy row(s)/col(s) are assumed to be zero. While implementing the solution, the dummy row(s) or dummy column(s) will not have any assignment.

### 4.11.2 Hungarian Method

This is one of the methods used to solve an assignment problem. In the first phase, reduction in rows and columns are done. In the second phase, the solution is optimized through iteration. These steps are illustrated while solving the example for better understanding of the concept. Students can formulate their own algorithms, if necessary.

**Illustration 4.12**

The admission office of a well-known institute has four tasks which vary in their difficulty level to be performed by four experienced people. The time taken is as in Table 4.7. Propose an effective task allocation to each person that minimizes the total man-hours.

Tasks

| | | *A* | *B* | *C* | *D* |
|---|---|---|---|---|---|
| | *P* | 16 | 52 | 34 | 22 |
| People | *Q* | 26 | 56 | 8 | 52 |
| | *R* | 76 | 38 | 36 | 30 |
| | *S* | 38 | 52 | 48 | 20 |

**Solution**
The initial step is to identify the least elements in each row and subtract the least elements from all the other elements in that row, respectively.

**Table 4.8** Row subtraction

Tasks

| | | *A* | *B* | *C* | *D* |
|---|---|---|---|---|---|
| | *P* | 0 | 36 | 18 | 6 |
| People | *Q* | 18 | 48 | 0 | 44 |
| | *R* | 46 | 8 | 6 | 0 |
| | *S* | 18 | 32 | 28 | 0 |

Once the row subtraction table is obtained, identify the least elements in each column in the row subtraction table and subtract them from all the other elements in the respective columns. This is shown in Table 4.9 (we identify least elements of each column in Table 4.8 and subtract them from other elements in the respective columns to obtain Table 4.9).

**Table 4.9** Column subtraction

Tasks

| | | *A* | *B* | *C* | *D* |
|---|---|---|---|---|---|
| | *P* | 0 | 28 | 18 | 9 |
| People | *Q* | 18 | 40 | 0 | 46 |
| | *R* | 46 | 0 | 6 | 0 |
| | *S* | 18 | 24 | 28 | 0 |

Next, we proceed to optimization.

| | | Tasks | | | |
|---|---|---|---|---|---|
| | | *A* | *B* | *C* | *D* |
| People | *P* | [0] | 28 | 18 | 9 |
| | *Q* | 18 | 40 | [0] | 46 |
| | *R* | 46 | [0] | 6 | ~~0~~ (x) |
| | *S* | 18 | 24 | 28 | [0] |

Begin with the first row and examine for zeros, until a row containing one zero is found. When found, mark this zero with □, and cross out ($x$) all other zeros in the corresponding rows and columns.

Look at the previous table.

1. The first row contains one zero. Mark this with □. Examine the row (first) and that column in which the zero is marked (first column). If there are any zeros, cross out ($x$) them. Here, we do not have any zeros.
2. The second row also contains one zero only. We mark this by □. The second row and third column are examined (as this 0 is in third column). We do not find any zeros here.
3. The third row contains two zeros. Leave this untouched for the moment.
4. Go to row 4, which contains one zero. Mark this by □. Examine fourth row and fourth column. The fourth column has a zero, above the marked zero. So, we cross ($x$).
5. Now, we are left with one zero in row 3. We mark this by □, which gives optimal assignment.

Hence, the optimal assignment is given by

| People | Tasks | Man-hours |
|---|---|---|
| P | A | 16 |
| Q | C | 8 |
| R | B | 38 |
| S | D | 20 |

The man-hours shown above should be taken from the given problem.

(*Note* if two zeros are encountered in successive rows, then examine the columns that have one zero and follow the above procedure for columns. If all columns also contain two zeros, then arbitrarily to any one zero, and follow the above procedure from then on).

The total man-hours required to complete all the four tasks are 84.

## Illustration 4.13

Give the optimal assignment for the following cost matrix (in rupees).

Tasks

| | | *I* | *J* | *K* | *L* |
|---|---|---|---|---|---|
| | *A* | 33 | 51 | 24 | 48 |
| People | *B* | 27 | 21 | 36 | 30 |
| | *C* | 39 | 48 | 45 | 36 |
| | *D* | 42 | 30 | 36 | 33 |

## Solution

**Table 4.10** Row subtraction

| | | | |
|---|---|---|---|
| 9 | 27 | 0 | 24 |
| 6 | 0 | 15 | 9 |
| 3 | 12 | 9 | 0 |
| 12 | 0 | 6 | 3 |

**Table 4.11** Column subtraction

| | | | |
|---|---|---|---|
| 6 | 27 | 0 | 24 |
| 3 | 0 | 15 | 9 |
| 0 | 12 | 27 | 0 |
| 12 | 0 | 6 | 1 |

We rewrite Table 4.11 to give assignments.

| | *I* | *J* | *K* | *L* | |
|---|---|---|---|---|---|
| *A* | 6 | 27 | [0] | 24 | |
| *B* | 3 | [0] | 15 | 9 | ✓ |
| *C* | [0] | 12 | 27 | 0 | |
| *D* | 12 | 0 | 6 | 1 | ✓ |
| | | ✓ | | | |

We follow the procedure as in illustration 4.12 and find row 4 has no assignment. Now look at the previous table.

1. We tick (✔) this row, i.e., row 4.
2. Then, we tick the columns that have zeros in row 4, i.e., the second column.
3. In this ticked column look for any assignment. If yes, tick the row which has assignment. Here, it is second row.
4. Now, draw straight lines through unticked rows (rows 1 and 3) and ticked columns (column 2).
5. In the rest of the remaining elements, identify the least element and subtract that element from the remaining elements, and add this least element at the intersections of the lines. Here, the least element is zero. So, we subtract this form from other elements as shown in the previous table and again proceed with the assignment procedure.

| | *I* | *J* | *K* | *L* |
|---|---|---|---|---|
| *A* | 6 | 30 | [0] | 24 |
| *B* | [0] | ~~0~~ | 12 | 6 |
| *C* | ~~0~~ | 15 | 9 | [0] |
| *D* | 6 | [0] | 3 | ~~0~~ |

The optimal assignment is

| People | | Tasks | | Costs (Rs.) |
|---|---|---|---|---|
| *A* | → | *J* | → | 24 |
| *B* | → | *I* | → | 27 |
| *C* | → | *L* | → | 36 |
| *D* | → | *K* | → | 30 |

The minimum cost is Rs. 117.

### 4.11.3 *Variations in Assignment Problem (Maximal Adjustment Problem)*

When in an assignment problem the objective is to maximize (instead of minimization), e.g., assign jobs in such a manner that profit is maximized, then it must first be converted to a minimization problem and solved in the usual way. This can be done by subtracting from the highest element all the elements of the given profit matrix; or equivalently, by placing negative (−ve) sign before each element of the profit matrix to make it cost matrix.

We will solve a few problems to make this clear.

**Illustration 4.14**

An organization producing four different products, viz. *A*, *B*, *C*, and *D* having four operators, viz. *P*, *Q*, *R*, and *S* who are capable of producing any of the four products, works effectively for 7 h/day. The time (in minutes) required for each operator for producing each of the products is given in the cells of the following matrix along with profit (Rs./unit).

| | Product | | | |
|---|---|---|---|---|
| Operator | *A* | *B* | *C* | *D* |
| *P* | 6 | 10 | 14 | 12 |
| *Q* | 7 | 5 | 3 | 4 |
| *R* | 6 | 7 | 10 | 10 |
| *S* | 20 | 10 | 15 | 15 |
| Profit (Rs./unit) | 3 | 2 | 4 | 1 |

Find the optimal assignment of operators that maximizes the profit.

**Solution**

This situation requires some analyses. We have to obtain an optimal allocation that maximizes the profit. We know that every operator works effectively for 7 h/day and also the times from above table.

First, we obtain the production matrix and from the production matrix, we obtain the profit matrix.

**Table 4.12** Production matrix

| Operator | Product | | | |
|---|---|---|---|---|
| ↓ | *A* | *B* | *C* | *D* |
| *P* | 70 | 42 | 30 | 35 |
| *Q* | 60 | 84 | 140 | 15 |
| *R* | 70 | 60 | 42 | 42 |
| *S* | 20 | 42 | 28 | 28 |

**Table 4.13** Profit matrix

| Operator ↓ | Product A | B | C | D |
|---|---|---|---|---|
| *P* | 210 | 84 | 120 | 35 |
| *Q* | 180 | 168 | 560 | 105 |
| *R* | 210 | 120 | 168 | 42 |
| *S* | 60 | 84 | 112 | 28 |

**Production Matrix Calculations**
Given: Number of hours/day = 7

$$\therefore \text{ Value is, production matrix } = \frac{7 \times 60}{6} = \frac{420}{6} = 70$$

where 6 is the number obtained from data, i.e., table under Illustration 4.14.

The rest of the values are calculated in a similar way. These values are multiplied by the profit/unit values in table (under Illustration 4.14) to get the profit matrix, as shown in Table 4.13.

Now, for assignment, it is necessary to convert the profit matrix to loss matrix. We do this by subtracting all elements of the matrix by the highest element, i.e., 560.

| Operator ↓ | Product A | B | C | D |
|---|---|---|---|---|
| *P* | 350 | 476 | 440 | 525 |
| *Q* | 380 | 392 | 0 | 455 |
| *R* | 350 | 440 | 392 | 518 |
| *S* | 500 | 476 | 448 | 532 |

Further, we perform row and column subtraction as usual.

**Table 4.14** Row subtraction

| Operator ↓ | Product A | B | C | D |
|---|---|---|---|---|
| *P* | 0 | 126 | 90 | 175 |
| *Q* | 380 | 392 | 0 | 455 |
| *R* | 0 | 90 | 42 | 168 |
| *S* | 52 | 28 | 0 | 84 |

**Table 4.15** Column subtraction

| Operator ↓ | Product A | B | C | D |
|---|---|---|---|---|
| P | 0 | 98 | 90 | 91 |
| Q | 380 | 364 | 0 | 371 |
| R | 0 | 62 | 42 | 84 |
| S | 52 | 0 | 0 | 0 |

Redrawing Table 4.15.

| Operator ↓ | Product A | B | C | D |
|---|---|---|---|---|
| P | [0] | 98 | 90 | 91 |
| Q | 380 | 364 | [0] | 371 |
| R | ~~0~~ | 62 | 42 | 84 |
| S | 52 | [0] | ~~0~~ | ~~0~~ |

In above table, one row is left without assignment. Hence, we move further,

| Operator ↓ | Product A | B | C | D |
|---|---|---|---|---|
| P | [0] | 36 | 90 | 29 |
| Q | 380 | 302 | [0] | 309 |
| R | ~~0~~ | [0] | 42 | 22 |
| S | 114 | ~~0~~ | 62 | [0] |

The optimum allocation is

| Operator | Product | Profit (Rs.) |
|---|---|---|
| *P* | *A* | 210 |
| *Q* | *C* | 560 |
| *R* | *B* | 120 |
| *S* | *D* | 28 |

Total Profit = Rs. 918.

**Illustration 4.15**

Given the cost and revenue data, obtain the optimal assignment that maximizes the profit.

**Table 4.16** Production cost

| | | X | Y | Z | W |
|---|---|---|---|---|---|
| | A | 98 | 120 | 90 | 122 |
| Plants | B | 110 | 126 | 90 | 138 |
| | C | 108 | 124 | 98 | 106 |
| | D | 110 | 128 | 96 | 132 |

**Table 4.17** Revenue

| | | X | Y | Z | W |
|---|---|---|---|---|---|
| | A | 100 | 136 | 98 | 124 |
| Plants | B | 120 | 140 | 102 | 148 |
| | C | 110 | 134 | 106 | 120 |
| | D | 116 | 130 | 108 | 138 |

**Solution**

We obtain the profit matrix in table below with the formula Profit = Revenue – Cost.

| | X | Y | Z | W |
|---|---|---|---|---|
| A | 2 | 16 | 8 | 2 |
| B | 10 | 14 | 12 | 10 |
| C | 2 | 10 | 8 | 14 |
| D | 6 | 2 | 12 | 6 |

| | X | Y | Z | W |
|---|---|---|---|---|
| A | 14 | 0 | 8 | 14 |
| B | 6 | 2 | 4 | 6 |
| C | 74 | 6 | 8 | 2 |
| D | 10 | 14 | 4 | 10 |

Above table gives the loss matrix. Now, we perform row and column subtraction.

**Table 4.18** Row subtraction

| | | | |
|---|---|---|---|
| 14 | 0 | 8 | 14 |
| 4 | 0 | 2 | 4 |
| 12 | 4 | 6 | 0 |
| 6 | 10 | 0 | 6 |

**Table 4.19** Column subtraction

| | | | |
|---|---|---|---|
| 10 | 0 | 8 | 14 |
| 0 | 0 | 2 | 4 |
| 8 | 4 | 6 | 0 |
| 2 | 10 | 0 | 6 |

| | *X* | *Y* | *Z* | *W* |
|---|---|---|---|---|
| *A* | 10 | [0] | 8 | 14 |
| *B* | [0] | X | 2 | 4 |
| *C* | 8 | 4 | 6 | [0] |
| *D* | 2 | 10 | [0] | 6 |

Above table gives the optimal assignment. Hence, the same can be represented as

| Plant | Product | Profit |
|---|---|---|
| *A* | *Y* | 16 |
| *B* | *X* | 10 |
| *C* | *W* | 14 |
| *D* | *Z* | 12 |

Total Profit = Rs. 52

### *4.11.4 Restrictions in Assignment*

Sometimes in an assignment, due to technical, legal, or other restrictions, it may not be possible to assign a particular facility to a particular job. This situation can be overcome by assigning a very high cost (say infinite cost) to the corresponding cell so that the activity will be automatically excluded from the optimal solution. We will illustrate this by an example.

**Illustration 4.16**
WELL DONE company has taken the third floor of a multistoried building for rent with a view to locate one of their zonal offices. There are five main rooms in this floor to be assigned to five managers. Each room has its own advantages and disadvantages, some have windows, some are closer to washrooms, or canteen, or secretarial pool. The rooms are of all different sizes and shape. Each of the five managers was asked to rank their room preferences among the rooms 301, 302, 303, 304, and 305. Their preferences were recorded in a table as indicated in below.

| $M_1$ | $M_2$ | $M_3$ | $M_4$ | $M_5$ |
|---|---|---|---|---|
| 302 | 302 | 303 | 302 | 301 |
| 303 | 304 | 301 | 305 | 302 |
| 304 | 305 | 304 | 304 | 304 |
| | 301 | 305 | 303 | |
| | | 302 | | |

Most of the managers did not list all the five rooms since they were not satisfied with some of the rooms, and they did not mention those rooms in the list. Assuming that their preferences can be quantified by numbers, find out as to which manager should be assigned to which room so that their total preference ranking is minimum.

**Solution**
We can first formulate the above problem according to the preference ranking in below table.

| | | $M_1$ | $M_2$ | $M_3$ | $M_4$ | $M_5$ | → Managers |
|---|---|---|---|---|---|---|---|
| | 301 | - | 4 | 2 | - | 1 | |
| | 302 | 1 | 1 | 5 | 1 | 2 | |
| Room | 303 | 2 | - | 1 | 4 | - | |
| No. | 304 | 3 | 2 | 3 | 3 | 3 | |
| | 305 | - | 3 | 4 | 2 | - | |

**Table 4.20** Row subtraction

| | | | | |
|---|---|---|---|---|
| - | 3 | 1 | - | 0 |
| 0 | 0 | 4 | 0 | 1 |
| 1 | - | 0 | 3 | - |
| 1 | 0 | 1 | 1 | 1 |
| - | 1 | 2 | 0 | - |

Since the row subtraction matrix contains zeros in all columns, the column subtraction matrix also is the same as row subtraction matrix.
Now, we assign rooms to managers.

| | $M_1$ | $M_2$ | $M_3$ | $M_4$ | $M_5$ |
|---|---|---|---|---|---|
| 301 | - | 3 | 1 | - | [0] |
| 302 | [0] | ~~0~~ | 4 | 0 | 1 |
| 303 | 1 | - | [0] | 3 | - |
| 304 | 1 | [0] | 1 | 1 | 1 |
| 305 | - | 1 | 2 | [0] | - |

The optimal assignment is

| Room No. | Manager | Rank |
|---|---|---|
| 301 | $M_5$ | 1 |
| 302 | $M_1$ | 2 |
| 303 | $M_3$ | 1 |
| 304 | $M_2$ | 2 |
| 305 | $M_4$ | 1 |

The total minimum ranking $= 1 + 2 + 1 + 2 + 1 = 7$.

### *4.11.5 Sensitivity in Assignment Problems*

The structure of an assignment problem leaves no scope for sensitivity analysis. Modest changes like one person able to do few jobs can be considered by repeating the man's row and adding a dummy column to square up the matrix. Addition of a constant to a row or a column also does not make any difference to the optimal assignment. Hence, there is no scope for altering the level of an assignment.

### 4.11.6 Traveling Salesman Problem

In a traveling salesman problem, there will be $n$ cities that a salesman must visit. The data would provide the distance between any two pairs of cities. The salesman should start from his home town/city and must visit all the cities at least once before arriving at the starting point or his home town. It is not required to follow a systematic order for visiting these cities; any order may be followed. The solution for the traveling salesman problem is to find the best visiting combination of the cities from the starting city that reduces the total cost of traveling for the company.

**Illustration 4.17**
A traveling salesman must visit five cities. It is required for him to start at a particular point (city) and visit each city once and return to the starting city. Given in below table is the traveling cost between cities.

| | 1 | 2 | 3 | 4 | 5 |
|---|---|---|---|---|---|
| 1 | - | 8 | 14 | 6 | 8 |
| 2 | 8 | - | 14 | 6 | 8 |
| 3 | 14 | 12 | - | 14 | 10 |
| 4 | 6 | 6 | 12 | - | 14 |
| 5 | 8 | 8 | 10 | 14 | - |

**Solution**
We proceed in the usual manner,

**Table 4.21** Row subtraction

| | | | | |
|---|---|---|---|---|
| - | 2 | 8 | 0 | 2 |
| 2 | - | 8 | 0 | 2 |
| 4 | 2 | - | 4 | 0 |
| 0 | 0 | 6 | - | 2 |
| 0 | 0 | 2 | 6 | - |

**Table 4.22** Column subtraction

| | | | | |
|---|---|---|---|---|
| - | 2 | 6 | 0 | 2 |
| 2 | - | 6 | 0 | 2 |
| 4 | 2 | - | 4 | 0 |
| 0 | 0 | 4 | - | 2 |
| 0 | 0 | 0 | 6 | - |

We rewrite Table 4.22 to facilitate assignments.

| | 1 | 2 | 3 | 4 | 5 | |
|---|---|---|---|---|---|---|
| 1 | - | 2 | 6 | [0] | 2 | ✓ |
| 2 | 2 | - | 6 | ~~0~~ | 2 | ✓ |
| 3 | 4 | 2 | - | 4 | [0] | |
| 4 | [0] | ~~0~~ | 4 | - | 2 | |
| 5 | ~~0~~ | [0] | 0 | 6 | - | |
| | | | | ✓ | | |

| | 1 | 2 | 3 | 4 | 5 |
|---|---|---|---|---|---|
| 1 | - | [0] | 4 | ~~0~~ | ~~0~~ |
| 2 | ~~0~~ | - | 4 | [0] | ~~0~~ |
| 3 | 4 | 2 | - | 6 | 0 |
| 4 | [0] | ~~0~~ | 4 | - | 2 |
| 5 | ~~0~~ | ~~0~~ | [0] | 8 | - |

We see in the above table, all rows have an assignment; therefore, we draw the following chart to analyze the route the traveling salesman has taken:

| From | To |
|---|---|
| 1 | 2 |
| 2 | 4 |
| 4 | 1 |
| 4 | - |
| 5 | - |

We see that the salesman has not visited cities 3 and 5. Therefore, the solution obtained is not optimal. So we rewrite the assignment matrix. In this table, we give the first assignment to the next least cost element (irrespective of its position) in the matrix and cross out ($x$) zeros in respective row and column. The further assignment is given to zeros as usual. In the assignment matrix, the next least cost element is 2.

| | 1 | 2 | 3 | 4 | 5 |
|---|---|---|---|---|---|
| 1 | - | ~~0~~ | 4 | ~~0~~ | [0] |
| 2 | ~~0~~ | - | 4 | [0] | ~~0~~ |
| 3 | 4 | [2] | - | 6 | ~~0~~ |
| 4 | [0] | ~~0~~ | 4 | - | 2 |
| 5 | ~~0~~ | ~~0~~ | [0] | 8 | - |

Now, we again draw the route chart.

| From | To |
|---|---|
| 1 | 5 |
| 5 | 3 |
| 3 | 2 |
| 2 | 4 |
| 4 | 1 |

Now, we see that the salesman has come back to the starting point after visiting all cities. So the above is the optimal assignment.

**Illustration 4.18**

Solve the traveling salesman problem, given the following data $C_{12} = 20$, $C_{13} = 14$, $C_{14} = 10$, $C_{23} = 5$, $C_{34} = 6$, $C_{25} = 10$, $C_{35} = 6$, and $C_{45} = 20$ where $C_{ij} = C_{ji}$, and there is no route between cities $i$ and $j$ if the value of $c_{ij}$ is not shown.

**Solution**

We draw the table for the above data.

| | 1 | 2 | 3 | 4 | 5 |
|---|---|---|---|---|---|
| 1 | - | 20 | 4 | 10 | - |
| 2 | 20 | - | 5 | - | 10 |
| 3 | 4 | 5 | - | 6 | 6 |
| 4 | 10 | - | 6 | - | 20 |
| 5 | - | 10 | 6 | 20 | - |

**Table 4.23** Row subtraction

| | | | | |
|---|---|---|---|---|
| - | 16 | 0 | 6 | - |
| 15 | - | 0 | - | 5 |
| 0 | 1 | - | 2 | 2 |
| 4 | - | 0 | - | 14 |
| - | 4 | 0 | 14 | - |

**Table 4.24** Column subtraction

| | | | | |
|---|---|---|---|---|
| - | 15 | 0 | 4 | - |
| 15 | - | 0 | - | 3 |
| 0 | 0 | - | 0 | 0 |
| 4 | - | 0 | - | 12 |
| - | 3 | 0 | 12 | - |

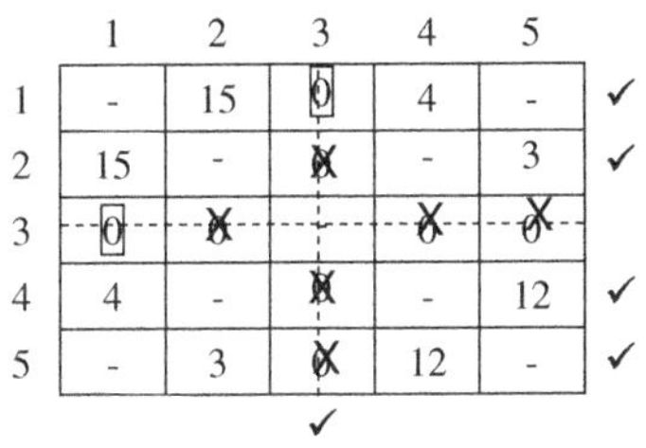


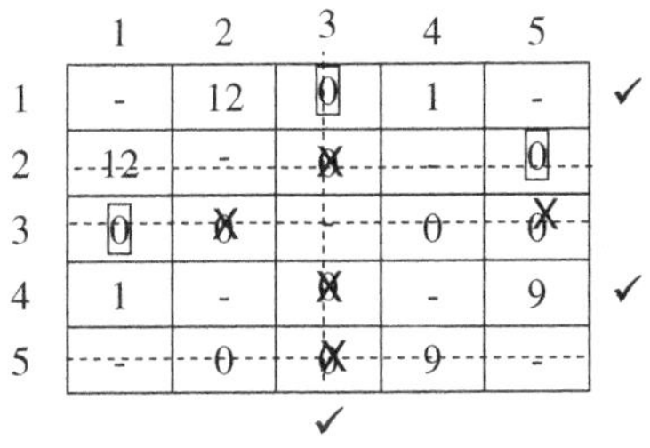

| | 1 | 2 | 3 | 4 | 5 |
|---|---|---|---|---|---|
| 1 | - | 1 | ~~0~~ | [0] | - |
| 2 | 12 | - | 1 | - | [0] |
| 3 | [0] | ~~0~~ | - | ~~0~~ | ~~0~~ |
| 4 | ~~0~~ | - | [0] | - | 8 |
| 5 | - | ~~0~~ | 1 | 8 | - |

Since every row in above table has assignment, we draw the route chart.

| From | To |
|---|---|
| 1 | 4 |
| 4 | 3 |
| 3 | 1 |

The salesman has not visited cities 2 and 5. Hence, the above solution is not optimal. So, we give the first allocation to next to least cost element, i.e., in cell $X_{23}$.

| | 1 | 2 | 3 | 4 | 5 |
|---|---|---|---|---|---|
| 1 | - | 11 | ~~0~~ | [0] | - |
| 2 | 12 | - | [1] | - | ~~0~~ |
| 3 | ~~0~~ | ~~0~~ | - | ~~0~~ | [0] |
| 4 | [0] | - | ~~0~~ | - | 8 |
| 5 | - | [0] | 1 | 8 | - |

All the rows in above table have assignments. Hence, we again draw the route chart to analyze.

| From | To |
|---|---|
| 1 | 4 |
| 4 | 1 |

Here, again the solution is not optimal as the salesman has not visited cities 2, 3, and 5. Now, we rewrite the above table, give the first assignment to next to least cost element, i.e., 1 (cell $X_{23}$), and the second assignment to next to next to least cost element, i.e., 8 (cell $X_{45}$), and then proceed as our usual assignment steps.

| | 1 | 2 | 3 | 4 | 5 |
|---|---|---|---|---|---|
| 1 | - | 11 | ~~0~~ | [0] | - |
| 2 | 12 | - | [1] | - | ~~0~~ |
| 3 | [0] | ~~0~~ | - | ~~0~~ | ~~0~~ |
| 4 | ~~0~~ | - | ~~0~~ | - | [8] |
| 5 | - | [0] | 1 | 8 | - |

Now, we draw the route chart again.

| From | To |
|---|---|
| 1 | 4 |
| 4 | 5 |
| 5 | 2 |
| 2 | 3 |
| 3 | 1 |

We can see that the sales man has visited all cities and returned to the starting point. Hence the above is the optimal solution.

### *4.11.7 Unbalanced Assignment Problem*

If the given assignment matrix is not a square matrix, then it is an unbalanced assignment problem. Addition of required number of dummy rows or columns with zero cost figures will balance the matrix. We solve an example to make this concept clear.

**Illustration 4.19**
Determine the optimal assignment of workers to jobs which results in maximum profits. Which job is to be refused or declined?

| | | Jobs | | | | |
|---|---|---|---|---|---|---|
| | | $J_1$ | $J_2$ | $J_3$ | $J_4$ | $J_5$ |
| | $W_1$ | 124 | 156 | 100 | 202 | 164 |
| Workers | $W_2$ | 142 | 168 | 122 | 146 | 118 |
| | $W_3$ | 174 | 184 | 222 | 142 | 162 |
| | $W_4$ | 96 | 128 | 174 | 154 | 160 |

**Solution**

We can see that the given matrix is a profit matrix and, more so, not a square matrix. So, we first convert it to a loss matrix and add a dummy row.

**Table 4.25** Loss matrix

| | | Jobs | | | | |
|---|---|---|---|---|---|---|
| | | $J_1$ | $J_2$ | $J_3$ | $J_4$ | $J_5$ |
| | $W_1$ | 98 | 66 | 122 | 20 | 58 |
| Workers | $W_2$ | 80 | 54 | 100 | 76 | 104 |
| | $W_3$ | 48 | 38 | 0 | 80 | 60 |
| | $W_4$ | 126 | 94 | 48 | 68 | 62 |
| | $W_5$ | 0 | 0 | 0 | 0 | 0 |

**Table 4.26** Row/Column subtraction

| | | | | |
|---|---|---|---|---|
| 78 | 46 | 102 | 0 | 38 |
| 26 | 0 | 46 | 22 | 50 |
| 48 | 38 | 0 | 80 | 60 |
| 78 | 46 | 0 | 20 | 14 |
| 0 | 0 | 0 | 0 | 0 |

Jobs

| | | $J_1$ | $J_2$ | $J_3$ | $J_4$ | $J_5$ | |
|---|---|---|---|---|---|---|---|
| | $W_1$ | 78 | 46 | 102 | [0] | 38 | |
| | $W_2$ | 26 | [0] | 46 | 22 | 50 | |
| Workers | $W_3$ | 48 | 38 | [0] | 80 | 60 | ✓ |
| | $W_4$ | 78 | 46 | ~~0~~ | 20 | 14 | ✓ |
| | $W_5$ | [0] | ~~0~~ | ~~0~~ | ~~0~~ | 0 | |
| | | | | ✓ | | | |

Jobs

| | | $J_1$ | $J_2$ | $J_3$ | $J_4$ | $J_5$ | |
|---|---|---|---|---|---|---|---|
| | $W_1$ | 78 | 56 | 116 | [0] | 38 | |
| | $W_2$ | 26 | [0] | 60 | 22 | 50 | |
| Workers | $W_3$ | 34 | 24 | [0] | 66 | 46 | ✓ |
| | $W_4$ | 64 | 32 | ~~0~~ | 6 | 0 | ✓ |
| | $W_5$ | [0] | ~~0~~ | 14 | ~~0~~ | 0 | |

## *4.11.8 Airline Crew Problem*

In the airline crew problem, it is required to find an optimal assignment for the airlines under specified conditions. These conditions may be with respect to the layover time or the rest period, the operations, etc. The following example makes this clear.

**Illustration 4.20**

Below table shows an airline timetable. It is required to obtain the pairing of flights that minimize layover time away from home. A condition is that a minimum layover/rest period of 4.75 or $4\frac{3}{4}$ h for crew is to be allowed. The airline operates throughout the week, i.e., 7 days. Find where the crew should be based (lesser layover time).

| Bangalore–Hyderabad | | | Hyderabad–Bangalore | | |
|---|---|---|---|---|---|
| Flight No. | Departure | Arrival | Flight No. | Departure | Arrival |
| 1 | 6 AM | 7 AM | 701 | 7 AM | 8.15 AM |
| 2 | 7 AM | 8 AM | 703 | 7.15 AM | 8.30 AM |
| 3 | 2.30 AM | 3.45 PM | 705 | 12.30 PM | 1.45 PM |
| 4 | 4.45 AM | 6.00 PM | 707 | 5.30 PM | 6.45 PM |

Construct layover time between flights for

(a) Crew based at Bangalore
(b) Crew based at Hyderabad.

**Solution**
Layover time between flights (min), when

| Flight No. | Crew based at Hyderabad | | | | Crew based at Bangalore | | | |
|---|---|---|---|---|---|---|---|---|
| | 701 | 703 | 705 | 707 | 701 | 703 | 705 | 707 |
| 1 | 1440 | 1470 | 330 | 630 | 1305 | 1290 | 1035 | 675 |
| 2 | 1380 | 1395 | 1710 | 570 | 1365 | 1350 | 1095 | 735 |
| 3 | 915 | 930 | 1245 | 1545 | 375 | 360 | 1485 | 1065 |
| 4 | 660 | 675 | 1110 | 1410 | 510 | 495 | 1620 | 1320 |

Construction of above table.

1. For crew based at Hyderabad, measure the rest period (time between arrival at Hyderabad and departure from Hyderabad), and fill these values horizontally. In the first row, we have

   Under 701: 7 AM–7 AM = 24 h = 24 × 60 = 1,440 min (Flights 1 and 1)
   Under 703: 7 AM–7.15 AM = 24.15 h = 24.15 × 60 = 1,449 min (Flights 1 and 2)
   Under 705: 7 AM–12.30 PM = 5.5 h = 5.5 × 60 = 330 min (Flights 1 and 3)
   Under 707: 7 AM–5.30 PM = 10.5 h = 10.5 × 60 = 630 min (Flights 1 and 4).

All these are filled horizontally in row 1. Similarly, calculate all the other values and fix up accordingly.

2. For crew based at Bangalore, again measure the rest period (time between arrival at Bangalore and departure from Bangalore), and fill these values vertically, i.e., calculations for Flight no. 1 will have to be filled under first column, or under Flight no. 701 column. We show this as follows:

   Flights 1 and 1: 8.15 AM–6 AM = 21.75 h = 1,305 min
   Flights 1 and 2: 8.15 AM–7 AM = 22.75 h = 1,365 min
   Flights 1 and 3: 8.15 AM–2.30 PM = 6.25 h = 375 min
   Flights 1 and 4: 8.15 AM–4.45 PM = 8.5 h = 510 min.

   Similarly, fill up all columns. Next, we compare each value on one side of the table (i.e., crew based at Hyderabad), with corresponding value on the other side of the table (crew based at Bangalore), and select the least between the two values. This enables us to construct below table.

| | | Jobs | | | | | |
|---|---|---|---|---|---|---|---|
| | | $J_1$ | $J_2$ | $J_3$ | $J_4$ | $J_5$ | |
| Workers | $W_1$ | 78 | 56 | 116 | [0] | 38 | |
| | $W_2$ | 26 | [0] | 60 | 22 | 50 | |
| | $W_3$ | 34 | 24 | [0] | 66 | 46 | ✓ |
| | $W_4$ | 64 | 32 | ~~0~~ | 6 | 0 | ✓ |
| | $W_5$ | [0] | ~~0~~ | 14 | ~~0~~ | 0 | |

Values with $^x$ stand for values taken from crew based at Hyderabad.

Next is the usual assignment procedure.

**Table 4.27** Row subtraction

| | | | |
|---|---|---|---|
| 975 | 960 | 0 | 300 |
| 795 | 780 | 525 | 0 |
| 15 | 0 | 885 | 705 |
| 15 | 0 | 615 | 825 |

**Table 4.28** Column subtraction

| | | | |
|---|---|---|---|
| 960 | 960 | 0 | 300 |
| 780 | 780 | 525 | 0 |
| 0 | 0 | 885 | 705 |
| 0 | 0 | 615 | 825 |

| | | 701 | 703 | 705 | 707 |
|---|---|---|---|---|---|
| Flight No. | 1 | 960 | 960 | [0] | 300 |
| | 2 | 780 | 780 | 525 | [0] |
| | 3 | [0] | ~~0~~ | 885 | 705 |
| | 4 | ~~0~~ | [0] | 615 | 825 |

Above table gives the solution. The optimum allocation would be

| Flight No. | Flight No. | Base | Times |
|---|---|---|---|
| 1 | 705 | Bangalore | 330 |
| 2 | 707 | Bangalore | 570 |
| 3 | 701 | Hyderabad | 375 |
| 4 | 703 | Hyderabad | 495 |

The total layover time is 1,770 min.

## 4.12 Review Exercises

1. What is a transportation problem?
2. Formulate the transportation problem. Compare it with the assignment problem.
3. Prove that assignment problem is a special case of transportation problem.
4. With reference to a transportation problem, define the following terms:

   (i) Feasible solution
   (ii) Basic feasible solution
   (iii) Optimal solution
   (iv) Non-degenerate basic feasible solution.

5. Using the NWCR method, determine the IBFS for the following transportation problem:

| | | Destination (To) | | | | Availability |
|---|---|---|---|---|---|---|
| | | 1 | 2 | 3 | 4 | ↓ |
| Source (From) | $X$ | 26 | 22 | 30 | 40 | 4000 |
| | $Y$ | 34 | 28 | 24 | 26 | 12000 |
| | $Z$ | 18 | 18 | 15 | 12 | 14000 |
| Requirement | → | 6000 | 6000 | 8000 | 10000 | |

6. Determine an IBFS using

   (i) Vogel's Approximation Method)
   (ii) Row Minima Method

by considering the following transportation problem:

| | | Destination (To) | | | | Availability |
|---|---|---|---|---|---|---|
| | | 1 | 2 | 3 | 4 | ↓ |
| Source (From) | 1 | 21 | 16 | 15 | 13 | 11 |
| | 2 | 17 | 18 | 14 | 23 | 13 |
| | 3 | 32 | 27 | 18 | 41 | 19 |
| Requirement | → | 6 | 10 | 12 | 15 | |

7. Determine the IBFS to the following transportation problem using

   (i) Matrix Minima Method
   (ii) Vogel's Approximation Method

| | | Destination (To) | | | | Availability |
|---|---|---|---|---|---|---|
| | | $D_1$ | $D_2$ | $D_3$ | $D_4$ | ↓ |
| Source (From) | $O_1$ | 1 | 2 | 1 | 4 | 30 |
| | $O_2$ | 3 | 3 | 2 | 1 | 50 |
| | $O_3$ | 4 | 2 | 5 | 9 | 20 |
| Requirement | → | 20 | 40 | 30 | 10 | |

8. Draw a flowchart of the transportation algorithm.
9. Determine the IBFS to the following transportation problem using

   (i) North-West Corner Rule
   (ii) Vogel's Approximation Method.

| | | Destination (To) | | | | | Availability |
|---|---|---|---|---|---|---|---|
| | | $A$ | $B$ | $C$ | $D$ | $E$ | ↓ |
| Source (From) | $A$ | 8 | 44 | 40 | 12 | 28 | 16 |
| | $B$ | 4 | 16 | 28 | 8 | 4 | 32 |
| | $C$ | 12 | 36 | 16 | 32 | 48 | 36 |
| Requirement | → | 12 | 12 | 16 | 20 | 24 | |

10. For a transportation problem, explain what is meant by an optimality test. Give the procedure to obtain the optimal solution from the IBFS.
11. Find the optimum solution to the following transportation problem for which the cost, origin-availabilities, and destination-requirements are given below.

| | | Destination (To) | | | | | Availability |
|---|---|---|---|---|---|---|---|
| | | A | B | C | D | E | ↓ |
| Source (From) | 1 | 3 | 4 | 6 | 8 | 8 | 20 |
| | 2 | 2 | 10 | 1 | 5 | 30 | 30 |
| | 3 | 7 | 11 | 20 | 40 | 15 | 15 |
| | 4 | 2 | 1 | 9 | 14 | 18 | 13 |
| Requirement | → | 40 | 6 | 8 | 18 | 6 | |

12. Solve the following cost minimizing transportation problem:

| | | Destination (To) | | | | | | Availability |
|---|---|---|---|---|---|---|---|---|
| | | $D_1$ | $D_2$ | $D_3$ | $D_4$ | $D_5$ | $D_6$ | ↓ |
| Source (From) | $O_1$ | 2 | 1 | 3 | 3 | 2 | 5 | 50 |
| | $O_2$ | 3 | 2 | 2 | 4 | 3 | 4 | 40 |
| | $O_3$ | 3 | 5 | 4 | 2 | 4 | 1 | 60 |
| | $O_4$ | 4 | 2 | 2 | 1 | 2 | 2 | 30 |
| Requirement | → | 30 | 50 | 20 | 40 | 30 | 10 | |

13. A product is stored at three origins. These products are supplied to four dealers for selling. The requirements are 180, 140, 100, and 120, respectively. The capacities of the sources are 300, 80, and 160, respectively. Further data are given below.

$$X_{11} = 54 \quad X_{12} = 46 \quad X_{13} = 62 \quad X_{14} = 138$$
$$X_{21} = 20 \quad X_{22} = 90 \quad X_{23} = 80 \quad X_{24} = 64$$
$$X_{31} = 60 \quad X_{32} = 108 \quad X_{33} = 70 \quad X_{34} = 114$$

Obtain the optimum solution for transportation of products such that the total costs are minimum.

14. A manufacturer has distribution centers at $X$, $Y$, and $Z$. These centers have an availability of 40, 20, and 40 units of the product, respectively. The retail outlets are $A$, $B$, $C$, $D$, and $E$ and the requirement is 25, 10, 20, 30, and 15 units, respectively. The center-outlet transportation cost is given below.

| Distribution Centre | Retail outlets | | | | |
|---|---|---|---|---|---|
| | *A* | *B* | *C* | *D* | *E* |
| *X* | 55 | 30 | 40 | 50 | 50 |
| *Y* | 35 | 30 | 100 | 45 | 60 |
| *Z* | 40 | 60 | 95 | 35 | 30 |

Determine the optimal distribution which minimizes the transportation costs.

15. The purchase manager, Mr. Shah of SRTC must decide on the amounts of fuel to buy from three possible vendors. The corporation refuels the buses regularly at the four depots within the area of its operations. The three oil companies have said that they can furnish up to the following amounts of fuel during the month: 2,75,000 l by oil company-1, 1,50,000 l by oil company-2, and 6,60,000 l by oil company-3. The required amount of fuel is 1,10,000 l at depot-1, 20,000 l at depot-2, 3,30,000 l at depot-3, and 4,40,000 l at depot-4. When the transportation costs are added to the price bid per liter supplied, the combined cost per liter for fuel in rupees from each vendor servicing a specific depot is shown below.

| | Company-1 | Company-2 | Company-3 |
|---|---|---|---|
| Depot 1 | 5.00 | 4.75 | 4.25 |
| Depot 2 | 5.00 | 5.50 | 6.75 |
| Depot 3 | 4.50 | 6.00 | 5.00 |
| Depot 4 | 5.50 | 6.00 | 4.50 |

Determine the optimum transportation schedule.

16. Solve the following TP:

| | | Destination (To) | | | Availability |
|---|---|---|---|---|---|
| | | $D_1$ | $D_2$ | $D_3$ | ↓ |
| Source (From) | $O_1$ | 4 | 3 | 2 | 10 |
| | $O_2$ | 2 | 5 | 0 | 13 |
| | $O_3$ | 3 | 8 | 6 | 12 |
| Requirement | → | 8 | 5 | 4 | |

17. The following table gives unit cost of transportation of a certain material from supply points $A$, $B$, and $C$ to demand points 1, 2, 3, and 4. The requirements and availabilities are also given:

| | To | | | | |
|---|---|---|---|---|---|
| From | 1 | 2 | 3 | 4 | Supply |
| $A$ | 5 | 1 | 7 | 10 | 10 |
| $B$ | 6 | 4 | 6 | 5 | 80 |
| $C$ | 3 | 2 | 5 | 8 | 15 |
| Demand | 75 | 20 | 50 | 25 | |

Due to previous contractual obligations, a minimum of 5 units of material must be supplied from $A$ to 4. As there is more demand than supply, demands at some points cannot be satisfied, for which there are penalty costs which are given by 5, 3, 2, and 4 for demand points 1, 2, 3, and 4, respectively. Find the optimum schedule of transportation and associated costs. Is the solution unique?

18. What is the least time transportation problem? Where is it useful? Describe the solution procedure for such a problem.
19. What is transshipment problem? How can a transshipment be modeled as a transportation problem?
20. Given the following data, find the optimum transportation routes:

| | $S_1$ | $S_2$ | $D_1$ | $D_2$ | Availability ↓ |
|---|---|---|---|---|---|
| $S_1$ | 0 | 2 | 2 | 1 | 8 |
| $S_2$ | 1 | 0 | 2 | 3 | 3 |
| $D_1$ | 2 | 2 | 0 | 2 | |
| $D_2$ | 1 | 3 | 2 | 0 | |
| Requirement → | | | 7 | 4 | |

21. Show that the procedure of separating the minimum elements not covered by any line, from all uncovered elements and adding the same element to all the elements lying in an intersection of two lines results in a matrix with the same optimal assignments as the original matrix.
22. Explain the Hungarian method to solve an assignment problem.

23. A company has four machines on which to do three jobs. Each job can be assigned to one machine only. The cost of each job on each machine is shown below.

| | | Machine | | | |
|---|---|---|---|---|---|
| | | *W* | *X* | *Y* | *Z* |
| Job | *A* | 18 | 24 | 28 | 32 |
| | *B* | 8 | 13 | 17 | 19 |
| | *C* | 10 | 15 | 19 | 22 |

What are the job assignments that minimize the total costs?

24. Obtain an optimal job machine assignment that minimizes the total cost. Find the minimum total cost.

| | | Machine | | | | | |
|---|---|---|---|---|---|---|---|
| | | 1 | 2 | 3 | 4 | 5 | 6 |
| Job | 1 | 13 | 13 | 16 | 23 | 19 | 9 |
| | 2 | 11 | 19 | 26 | 16 | 17 | 18 |
| | 3 | 12 | 11 | 4 | 9 | 6 | 10 |
| | 4 | 7 | 15 | 9 | 14 | 14 | 13 |
| | 5 | 9 | 13 | 12 | 8 | 14 | 11 |

25. To simulate interest and provide an atmosphere for intelligence discussion, a finance faculty in a business school decides to hold seminars on four contemporary topics—leasing, portfolio management, private mutual funds, swap, and options. Such seminars should be held once in a week in the afternoons. However, scheduling these seminars (one for each topic, and not more than one seminar per afternoon) has to be done carefully so that the number of students unable to attend is kept to a minimum. A careful study indicates that the number of students who cannot attend a particular seminar on a specific day is as follows:

| | Leasing | Portfolio management | Privatemutual funds | Swap and options |
|---|---|---|---|---|
| Monday | 50 | 40 | 60 | 20 |
| Tuesday | 40 | 30 | 40 | 30 |
| Wednesday | 60 | 20 | 30 | 20 |
| Thursday | 30 | 30 | 20 | 30 |
| Friday | 10 | 20 | 10 | 30 |

Find an optimal schedule of these seminars. Also, find the total number of students who will be missing at least one seminar.

26. Four operators $O_1$, $O_2$, $O_3$, and $O_4$ are available to a manager who has got four jobs $J_1$, $J_2$, $J_3$, and $J_4$ done by assigning one job to each operator. Given are the times needed by different operators for different jobs.

    (i) How should the manager assign the jobs so that the total time needed for four jobs is minimum?
    (ii) If job $J_2$ is not to be assigned to operator $O_2$, what is the assignment and the additional total time required for the same?

| | $J_1$ | $J_2$ | $J_3$ | $J_4$ |
|---|---|---|---|---|
| $O_1$ | 12 | 10 | 10 | 8 |
| $O_2$ | 14 | 12 | 15 | 11 |
| $O_3$ | 6 | 10 | 16 | 4 |
| $O_4$ | 8 | 10 | 9 | 7 |

27. Imagine yourself to be the Executive Director of a 5-star hotel, which has four banquet halls that can be used for all functions, including weddings. The halls are all about the same size but the facilities in each hall differ. During a heavy marriage occasion, four parties approach you to reserve a hall for the marriage to be allocated on the same day. These parties are told that the first choice among the halls would cost Rs. 10,000 for the day. They were also to indicate the second, third, and fourth preferences, and the price they would be willing to pay. Parties *A* and *D* indicated that they would not be interested in halls 3 and 4. Other details are given below:

| Marriage party | Hall | | | |
|---|---|---|---|---|
| | 1 | 2 | 3 | 4 |
| *A* | 10,000 | 9,000 | - | - |
| *B* | 8,000 | 10,000 | 8,000 | 5,000 |
| *C* | 7,000 | 10,000 | 6,000 | 8,000 |
| *D* | 10,000 | 8,000 | - | - |

    Here (–) indicates that the party does not prefer that particular hall. Decide on an allocation that will maximize the revenue to your hotel.

28. Discuss in detail the traveling salesman problem.
29. ABC icecream company has a distribution depth in Greater Kailash Part I for distributing icecream in South Delhi. There are four vendors located in different parts of South Delhi (*A, B, C, D*), who have to be supplied icecream everyday. The following matrix describes the distances in (km) between the depot and four vendors:

To

| From | | Depot | Vendor A | Vendor B | Vendor C | Vendor D |
|---|---|---|---|---|---|---|
| | Depot | - | 3.5 | 3 | 4 | 2 |
| | Vendor A | 3.5 | - | 4 | 2.5 | 3 |
| From | Vendor B | 3 | 4 | - | 4.5 | 3.5 |
| | Vendor C | 4 | 2.5 | 4.5 | - | 4 |
| | Vendor D | 2 | 3 | 5.5 | 4 | - |

What route should the company follow so that the total distance traveled is minimized?

30. Solve the following TSP, given the costs of traveling. Determine the optimal route.

| | *A* | *B* | *C* | *D* | *E* |
|---|---|---|---|---|---|
| *A* | - | 4 | 8 | 14 | 2 |
| *B* | 10 | - | 4 | 16 | 4 |
| *C* | 14 | 12 | - | 8 | 12 |
| *D* | 20 | 6 | 10 | - | 8 |
| *E* | 2 | 4 | 4 | 16 | - |

31. A salesman has to visit five cities *A*, *B*, *C*, *D*, and *E*. The distances (in hundred miles) between the cities are as follows:

To

| From | | *A* | *B* | *C* | *D* | *E* |
|---|---|---|---|---|---|---|
| | *A* | - | 7 | 6 | 8 | 4 |
| | *B* | 7 | - | 8 | 5 | 6 |
| From | *C* | 6 | 8 | - | 9 | 7 |
| | *D* | 8 | 5 | 9 | - | 8 |
| | *E* | 4 | 6 | 7 | 8 | - |

32. (a) The solution to an assignment problem is inherently degenerate—Explain.

(b) An air transport corporation picks up and delivers freight where customers require. The company has two types of aircraft, $X$ and $Y$, with equal loading capacities but different operating costs as shown. Find the optimal assignment that minimizes the total cost.

| Types of Aircraft | Cost per km in Rs. | |
|---|---|---|
| | Empty | Loaded |
| $X$ | 1.00 | 2.00 |
| $Y$ | 2.00 | 3.00 |

33. The present four locations of the aircraft the company has are:

$$J \rightarrow X \quad K \rightarrow Y \quad L \rightarrow Y \quad M \rightarrow X$$

Four customers of the company located at $A$, $B$, $C$, and $D$ want to transport nearly the same load to their final destinations. The final destinations are at a distance of 600, 300, 100, and 500 km from loading points $A$, $B$, $C$, and $D$, respectively. Distance between aircraft locations and loading points are given below.

| | $A$ | $B$ | $C$ | $D$ |
|---|---|---|---|---|
| $J$ | 300 | 200 | 400 | 100 |
| $K$ | 300 | 100 | 300 | 300 |
| $L$ | 400 | 100 | 100 | 400 |
| $M$ | 200 | 200 | 300 | 200 |

Determine the optimum allocation and total cost.

# Chapter 5
# Network Models

The key takeaways from this chapter are listed below:

- Constructing a network chart or diagram from a given set of activity data points
- Identify the critical paths, floats, and slacks in the network
- Crashing of activities and altering the total duration of the project
- Learn about flows in the network through the maximal spanning tree algorithm.

The term network flow program describes a type of model that is a special case of the more general linear program. The class of network flow programs includes such problems as the transportation problem, the assignment problem (previous chapter), the shortest path problem, and maximum flow problem. It is an important class because many aspects of actual situations are readily recognized as networks and the representation of the model is much more compact than the general linear program. When a situation can be entirely modeled as a network, very efficient algorithms exist for the solution of the optimization problem, many times more efficient than linear programming in the utilization of computer time and space resources. In this chapter, we introduce the reader to important day-to-day techniques such as CPM and PERT and also to the maximum spanning tree algorithm. The illustrations demonstrate the methodologies from constructing a network diagram to calculating the various project times. Let us start by constructing a network.

A network is a set of nodes (vertices) and a set of arcs (edges). Each node represents a location (say town or city), and each arc represents the connection (road limit) between two different locations. The number on each arc represents the distance between the locations. There are three types of network techniques.

1. Project evaluation and review technique (PERT)
2. Critical path method (CPM)
3. Minimal spanning tree algorithm.

R. Srinivasan, *Strategic Business Decisions*,
DOI: 10.1007/978-81-322-1901-9_5, © Springer India 2014

## 5.1 Representation Using Network Diagram

In project scheduling, we have to first draw an arrow diagram showing interdependence and precedence relationships among activities. The basic definitions in a network representation of a project are given below.

**Activity**: An activity is defined as the operation that utilizes the resources. Every activity has a beginning and an ending.

An activity is represented by an arrow (⟶) which indicates the direction of progress of the project.

There are different kinds of activities one experiences in a network. The activities that have to complete before the start of the other activities are the **predecessor** activities. Activities whose execution is impossible before the completion of certain activities are called **successor** activities, i.e., they cannot be started before completing their respective predecessor activities. Some activities do not consume any resources in a project, but have to be accomplished to complete the project. These kinds of activities are called **dummy** activities, represented by dashed arrow (⋯⋯▶).

**Event**: Completion of an activity and start of the successive activities at any point in time are represented by an event or a node. Events are represented by circles ◯.

There are three different kinds of events in a network. One or more activities converge at a particular event, such events are called **merge** events. If one or more activities leave a particular event, such events are called **burst** events. At an event, one or more activities can join, and one or more activities can leave that particular event at the same time; such events are called **merge and burst** events.

### *5.1.1 Rules for Drawing a Network Diagram*

There are some rules to be remembered while drawing a network diagram. They are as follows:

1. As discussed earlier, the events are represented by arrows, and the activities are represented by events. The arrow length does not have any restrictions; it can be of any length.
2. One activity can be represented by only one arrow, starting and ending at a circle or an event.
3. The tail of the arrow signifies the start of the activity; its head represents the completion.
4. The initial event is the predecessor event for all the other activities; hence, this should be represented at the beginning. No two activities can be identified by an end event. The network representation should be done in such a way that the arrows do not cross and are only straight.
5. Dummy activities should be represented by dashed arrows.
6. Finally, the network should have only one start and end event, respectively.

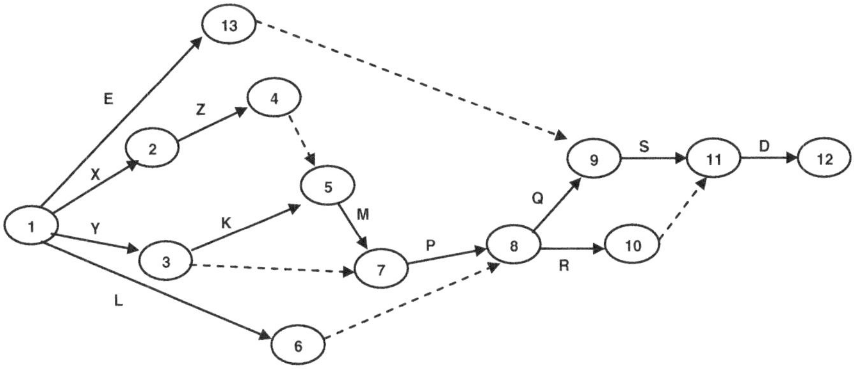


**Fig. 5.1**

## Illustration 5.1

Represent the following data by a network diagram.

| Activity | Immediate predecessor |
|---|---|
| *X* | – |
| *Y* | – |
| *Z* | *X* |
| *K* | *Y* |
| *M* | *Z*, *K* |
| *P* | *M*, *V* |
| *R* | *P*, *L* |
| *Q* | *R*, *P* |
| *T* | *M*, *Q* |
| *S* | *E*, *T* |
| *E* | *X* |
| *D* | *R*, *S* |

## Solution

The network is shown in Fig. 5.1. The activities with no precedence will have to be indicated first. Activities *Z*, *K* can be started only after completion of *X* and *Y*, respectively. Similarly, *M* can be started only after both *Z* and *K* are completed. The requirement for *P* is the completion of *M* and *Y* and so on. The → (arrow line) indicates the merging of two events/the completion of two events which ensures start of the next event. The numbers in the circles have no relevance when alphabets are used to represent activities.

## 5.2 PERT and CPM

PERT and CPM are two analytical project management techniques for planning, scheduling, and controlling of projects. Both were developed in the US between 1956 and 1958—CPM by Walker for scheduling and PERT by a team of engineers working in the US missile project.

PERT and CPM use time-oriented methods in that they lead to a time schedule for the project.

### *5.2.1 Applications*

Both can be used to solve a variety of problems such as

a. Dam construction
b. Highway construction
c. Building construction
d. Space flight
e. Design of machine prototype.

### *5.2.2 Phases in PERT/CPM*

There are four phases in PERT/CPM techniques. The first phase is the **planning** phase where the entire project is subdivided into many small projects. These small projects are in turn divided into activities, followed by the relationship identification between various activities. The activities are grouped into critical, important, less important, etc.

The next phase is to **schedule** these activities, which involves indicating the start and finish of each activity, respectively, indicating the activities that fall on the critical path. Float times must also be indicated for the non-critical activities. This scheduling phase is followed by **resource allocation**. In this phase, all the finance, material-, men-, and machine-related issues are addressed. Care is taken such that there are no conflicts anywhere in the project resources.

The last phase is to control the project activities by applying project management techniques. Regular work reports and appraisals help in controlling the resources better. Immediate actions whenever required can be initiated by strict controlling and monitoring of the project.

### *5.2.3 Differences Between PERT and CPM*

Normally, PERT and CPM are managing and controlling of large and complex projects. They are managerial techniques used in analyzing relationships between different activities in a project network. However, there are few differences between the two techniques. They are given under:

1. PERT is a probabilistic model. There are uncertainties in the activity durations. Multiple time estimates are made to find the probability of completing the project within the scheduled time. On the contrary, CPM is a deterministic model with single time estimates. These time estimates could be based on past experiences.
2. In case of PERT, statistical analysis is used to estimate the probability of finishing an activity and also the entire project. No statistical analysis is used in case of CPM to determine the probability of finish of an activity and the entire project.
3. In case of PERT, the result of analysis is expressed in terms of events. So PERT is an event-oriented technique. In case of CPM, the result of analysis is expressed in terms of activities. So CPM is an activity-oriented technique.
4. Costs are not considered directly in case of PERT, but CPM considers costs of project scheduling under minimization in a direct manner.
5. PERT is used for non-repetitive jobs such as planning a spacecraft assembly. CPM is used for repetitive jobs such as planning the construction of a house.
6. PERT is applied for planning and scheduling research programs, whereas CPM is applied for solving business problems.

#### 5.2.3.1 Disadvantages of PERT and CPM

1. If the input provided is not clear, then the development of the network itself poses to be a problem in these techniques.
2. Sometimes, the variances in the results may be on the higher side, which could lead to increase in project costs and time. This could frustrate the top management.
3. Adroitness is required for implementation of these techniques. Unskilled labor only costs without any payback and also compromises a lot on the quality of work.
4. Personnel experience plays an important role in implementation of these techniques. The level of network detail can be judged by experienced personnel, thus ensuring that there are no hassles in the future execution of the project.

## 5.3 Critical Path Method

### *5.3.1 Time Estimates and Critical Path*

The main objective of time estimates is to aid in preparing a planning schedule of the project. It should consist of:

1. **Earliest activity start time**: It is that time before which a particular activity cannot be started.
2. **Latest activity start time**: This time beyond which the start of an activity cannot be delayed any more is the latest activity start time.
3. **Earliest Activity Finish Time**: This is the time before which a particular activity cannot be completed.
4. **Latest Activity Finish Time:** This is the time beyond which the completion of a particular activity cannot be extended.
5. **Floats and Slacks**

   (a) **Total Float (TF)**: The time by which an activity can be delayed without affecting the total project completion time is the total float for an activity.
   (b) **Free Float (FF)**: The portion of total float within which an activity can be manipulated, such that the floats of subsequent activities do not get affected, is the FF of an activity.
   (c) **Independent Float (IF)**: The portion of total float within which the start of an activity can be delayed, such that the floats of preceding activities are not affected, is the IF for an activity.

6. **Critical Path**: Critical path is the longest path in the network diagram. The activities that fall on the critical path are called critical activities. These activities have to be completed in the specified time duration only. Any delay in the completion of the critical activities directly affects the total project completion time.

**Illustration 5.3**

Draw the network for the following data and find the following:

1. Earliest start time (EST)
2. Latest start time (LST)
3. Earliest finish time (EFT)
4. Latest finish time (LFT)
5. TF, IF, and FF. Represent the critical path in the network and estimate the total project duration.

| Activity | Duration |
|---|---|
| 1–2 | 30 |
| 2–4 | 24 |
| 1–3 | 24 |
| 3–5 | 10 |
| 2–5 | 24 |
| 4–6 | 40 |
| 5–6 | 24 |
| 6–7 | 32 |

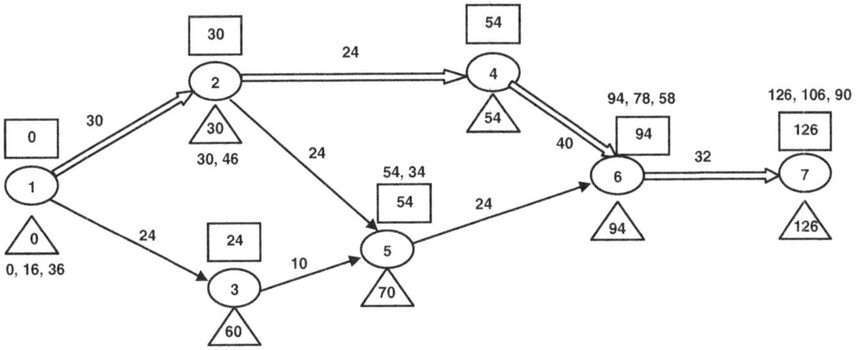


**Fig. 5.2**

**Solution**

The network is shown in Fig. 5.2, and the calculations are as in below table.

| Activity $i$–$j$ | Duration $D_{ij}$ | EST | EFT<br>$D_{ij}$ + EST | LST<br>LFT − $D_{ij}$ | LFT | TF | FF | IF |
|---|---|---|---|---|---|---|---|---|
| 1–2 | 30 | 0 | 30 | 0 | 30 | 0* | 0 | 0 |
| 2–4 | 24 | 30 | 54 | 30 | 54 | 0* | 0 | 0 |
| 1–3 | 24 | 0 | 24 | 36 | 60 | 36 | 0 | 0 |
| 3–5 | 10 | 24 | 34 | 60 | 70 | 36 | 20 | 56 |
| 2–5 | 24 | 30 | 54 | 46 | 70 | 16 | 0 | 0 |
| 4–6 | 40 | 54 | 94 | 54 | 94 | 0* | 0 | 0 |
| 5–6 | 24 | 54 | 74 | 70 | 94 | 20 | 20 | 36 |
| 6–7 | 32 | 94 | 126 | 94 | 126 | 0* | 0 | 0 |

The critical path is the path where closure lines are drawn, i.e.: 1–2–4–6–7. Any change in the durations of the activities on the critical path will affect the total project duration.

Procedure for calculating project duration:

1. The squares (□) and triangles (Δ) have to be drawn above and below, respectively, for each activity.
2. In the first square, i.e., above activity 1, the duration will be 0, similarly for the triangle below also. This is the same for all cases.
3. Now, in order to find the value in the square (□) for other events, we have to find the different paths that lead us to that activity.

   For example: For activity 2, we can reach 2 by one way only, i.e., 1–2. This takes 30 days, so the value in the (□) will be 30, whereas for 5, we have two routes, 1–2–5 or 1–3–5.

   Direction 1–2–5 gives 0 + 30 + 24 = 54

   Direction 1–3–5 gives 0 + 24 + 10 = 34

In such cases, we write down all obtained values from different routes on the top of the square (above □ of activity 5) and select the maximum value. This indicates activity 5 can be started only after 54 days from the start of the project, or the EST is calculated from the □ at the tail of the event.
In a similar way, we find the routes to reach activity 7 and we get the total project duration as 126 days.

4. Now, we have to fill up the triangles ($\Delta$). For this, we start from activity 7 (i.e., last activity). The values in these triangles indicate the LFT for a particular activity.
   The LFT for activity 7 will have to be taken down as 126 only, i.e., total project duration. Now, we go backwards.
   For 6, the value in_will be 126 − 32 = 94.
   For_of activity 5, we have route 7–6–5 = 126 − 32 − 24 = 70.
   For_of activity 2, we have 7–6–5–2 = 126 − 32 − 24 = 46.
   And we have 7–6–4–2 = 126 − 32 − 40 − 24 = 30.
   In such cases, we write down the obtained values below the triangle ($\Delta$) (of activity 2) and select the minimum. Here, it is 30. In the same way, we reach activity 1, and LFT for activity 1 is always 0.

*Note* For any activity, if the difference of □ and $\Delta$ is zero, then that activity falls on the critical path.

Now, we calculate all requirements of the question and tabulate as in the previous table.

1. The EST is the value in the square □ at the tail of an event.
   For example: for activity 1–2, we have value of □ as 0 at the tail of that activity.
   For activity 2–4, we have 30 as EST, and similarly for all activities, the ESTs are tabulated as in the previous table.
2. The LFT is the value in the triangles $\Delta$ at the head of an activity.
   For example: for activity 1–2, it is 30; for activity 1–3, it is 60.
3. The EFT is the sum of the duration ($D_{ij}$) and the EST of that respective activity as shown in the previous table.
4. The LST is the difference of the LFT and activity duration ($D_{ij}$) as in the previous table.
5. The total float is calculated by the formula:
   TF = LST − EST
   or
   TF = LFT − EFT
   The FF is calculated by the formula.

$$\text{Free Float (FF)} = \text{Total float} - \text{Head stack}$$

where
Head Stack = Difference between the value of the triangle and value of the square at the head of an event.
For 1–2, TF = 0,

$$\text{Head Stack} = 30 - 30 = 0\,(\text{at head of } 1{-}2, \text{i.e., at top of } 2)$$

$\therefore$ FF = 0 − 0 = 0.

6. IF = TF − Tail slack, where Tail Slack = Difference between the value of the triangle and value of the square at the tail of an event.
For 1–2, TF = 0,

$$\text{Tail slack} = 0 - 0 = 0\,(\text{an top of event } 1).$$

$\therefore$ IF = 0 − 0 = 0.
This is calculated in the same way for all the activities as shown in the previous table.

*Note* In the TF column, if the value is zero, then that activity falls on the critical path.

**Illustration 5.4**.

| Activity | Duration (days) |
|---|---|
| 1–2 | 6 |
| 1–4 | 4 |
| 1–5 | 4 |
| 2–3 | 8 |
| 4–7 | 8 |
| 4–6 | 14 |
| 5–6 | 8 |
| 3–8 | 4 |
| 7–8 | 10 |
| 6–9 | 12 |
| 8–9 | 6 |

Draw the network for the above data and find the fall EST, EFT, LST, LFT, TF, IF, FF.

**Solution**
The critical path is 1–4–6–9 and the total project duration is 30 days. All the calculations are made and tabulated in next table (Fig. 5.3).

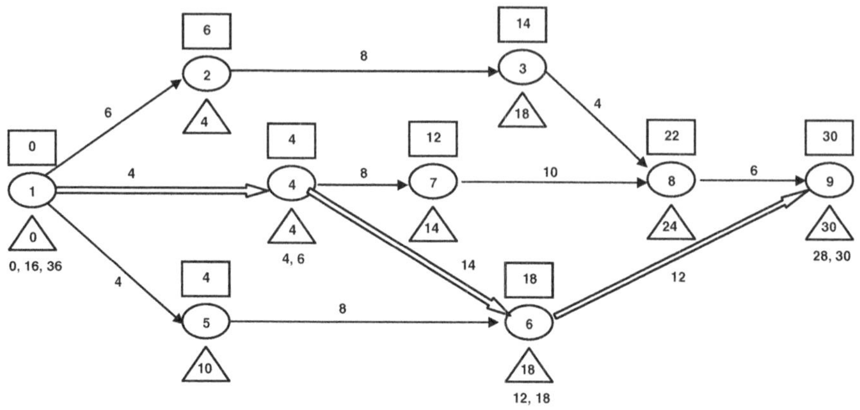


**Fig. 5.3**

| Activity $i$–$j$ | Duration $D_{ij}$ | EST | EFT $Dij$ + EST | LST LFT − $D_{ij}$ | LFT | TF | FF | IF |
|---|---|---|---|---|---|---|---|---|
| 1–2 | 6 | 0 | 6 | 6 | 12 | 6 | 0 | 0 |
| 1–4 | 4 | 0 | 4 | 0 | 4 | 0* | 0 | 0 |
| 1–5 | 4 | 0 | 4 | 6 | 10 | 6 | 0 | 0 |
| 2–3 | 8 | 6 | 14 | 12 | 20 | 6 | 0 | –6 |
| 4–7 | 8 | 4 | 12 | 6 | 14 | 2 | 0 | 0 |
| 4–6 | 14 | 4 | 18 | 4 | 18 | 0* | 0 | 0 |
| 5–6 | 8 | 4 | 12 | 10 | 18 | 6 | 6 | 0 |
| 3–8 | 4 | 14 | 18 | 20 | 24 | 6 | 4 | −2 |
| 7–8 | 10 | 12 | 22 | 14 | 24 | 2 | 0 | −2 |
| 6–9 | 12 | 18 | 30 | 18 | 30 | 0* | 0 | 0 |
| 8–9 | 6 | 22 | 28 | 24 | 30 | 2 | 2 | 0 |

*Note* In the TF Column, * above zeros indicates activities that fall on critical path

## *5.3.2 Project Evaluation and Review Technique*

PERT aids in completion of a particular project within a specified time. It takes uncertainties into account by associating three time values with an activity. They are as follows:

**The optimistic time**: The optimistic time is denoted by $t_o$. It is the shortest time within which a particular activity can be completed (Fig. 5.4).

**The most likely time**: The most likely time is denoted by $t_m$. It is the normal time taken to complete a particular activity, assuming that the project follows its normal planned course.

**The pessimistic time**: The pessimistic time is represented by $t_p$. The pessimistic time represents the maximum deviation with respect to completion of a

**Fig. 5.4**

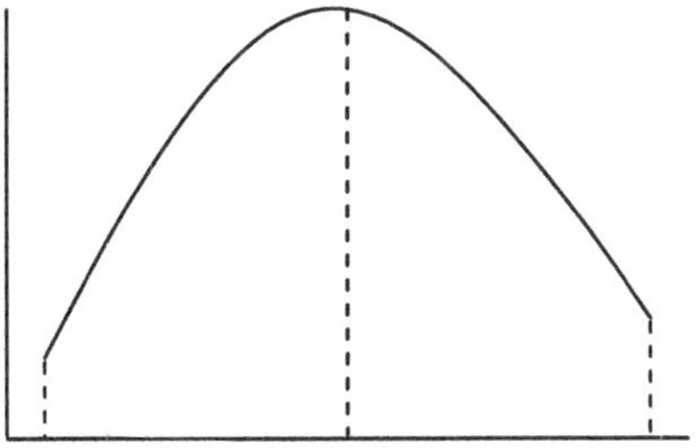


particular activity, or in other words, it is the longest duration taken to complete an activity assuming high variances in the project estimates.

**Expected time**: The average time taken to complete a particular activity on repeating it for a number of times is the expected time for an activity. The expected time for an activity follows beta distribution. It is represented by $t_e$ and calculated by the formula:

$$t_e = (t_o + 4t_m + t_p)/6$$

**Variance**: The variance of an activity is calculated by the formula:

$$\sigma^2 = [(t_p - t_o)/6]^2$$

**Estimation of probability**: the probability of completing a project can be estimated using the formula:

$$Z_{cal} = \frac{\text{Expected completion date} - \text{Project duration}}{\sqrt{\text{Project variance}}}$$

**Illustration 5.6**

Draw the network diagram for the following data and compute (1) variance, (2) critical path, (3) project length, and (4) probability so that the project will be completed in 51 days.

| Activity $i$–$j$ | $t_o$ In days | $t_p$ | $t_m$ |
|---|---|---|---|
| 1–2 | 6 | 6 | 6 |
| 2–3 | 6 | 18 | 12 |
| 2–4 | 4 | 12 | 8 |
| 3–5 | 8 | 14 | 12 |
| 5–7 | 6 | 10 | 8 |
| 5–6 | 0 | 0 | 0 |
| 6–7 | 4 | 16 | 10 |
| 4–6 | 8 | 16 | 12 |

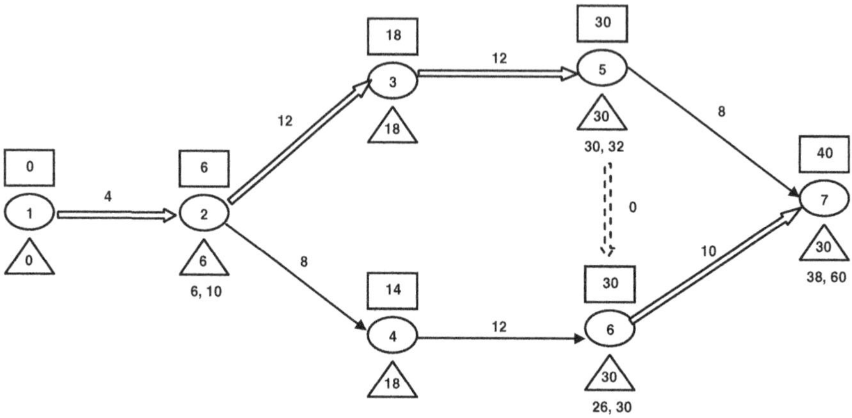


**Fig. 5.5**

**Solution**

The network is shown in Fig. 5.5.

The critical path is 1–2–3–5–6–7. Activity 5–6 is shown in dashed lines as this is a dummy activity, i.e., zero values.

The project duration or play at length is 40 days.

Below table gives the expected times and variances for all activities.

| Activity ($i$–$j$) | $t_o$ | $t_p$ Days | $t_m$ | Expected time ($T_e$) Days | Variance |
|---|---|---|---|---|---|
| 1–2 | 6 | 6 | 6 | 6 | 0* |
| 2–3 | 6 | 18 | 12 | 12 | 4* |
| 2–4 | 4 | 12 | 8 | 8 | 1.78 |
| 3–5 | 8 | 14 | 12 | 12 | 1* |
| 5–7 | 6 | 10 | 8 | 8 | 0.45 |
| 5–6 | 0 | 0 | 0 | 0 | 0* |
| 6–7 | 4 | 16 | 10 | 10 | 4* |
| 4–6 | 8 | 6 | 12 | 12 | 1.78 |

The expected time ($T_e$) is calculated from the formula $T_e = \frac{T_0 + 4T_m + T_p}{6}$ and the variance $\sigma^2 = \left\{\frac{T_p - T_0}{6}\right\}^2$

## 5.3.3 *Crashing*

By crashing an activity, we bring down the time taken to complete that respective activity. Normally, activities on the critical path are crashed, as only those activities influence the total project completion time. Reducing the time for these activities reduce the total project time.

#### 5.3.3.1 Elements of Crashing

The cost slope is calculated by the formula:

Cost Slope = (Crash cost − Normal cost)/(Normal time − Crash time)

By calculating the cost slope, the extra amount that is to be incurred per day for reducing the activity duration can be estimated.

The total cost of the project is calculated from the formula:

**Total Project Cost** = Direct Cost + Indirect Cost, where

**Direct cost** is the cost associated with the actual execution of an activity and

**Indirect cost** is the cost for implementing the project. A few examples of indirect cost are supervisors' expenses, managers' salary, etc. The illustration given below makes the concept clear.

**Illustration 5.8**

Table below gives the jobs, their normal time and cost, and crash time and cost for a project.

Indirect cost for the project is Rs. 300/day.

1. Draw the network of the project.
2. What is the normal duration and normal cost of the project?
3. Find the optimum duration and minimum project cost.
4. If all the activities are crashed, what will be the project duration and corresponding costs?

**Solution**

The network is as shown in Fig. 5.6.

Calculate the total expected project duration and the critical path.

The critical path is 1–2–3–4–6 and the total project duration is 20 days.

***Crashing***

By crashing some activities, we see that the project duration is reduced and also the corresponding costs. The expected project length can be reduced by crashing those activities that lie on the critical path.

In order to do that, we obtain a few figures from the given data: Normal Project Duration = 20 days

Normal Project Cost = Direct cost (Total Normal Cost) + Indirect cost

Total normal Cost = Rs. 9,200 (from next table, sum of normal costs)

| | Normal time | | Crash | |
|---|---|---|---|---|
| Job | Time in days | Cost in Rs. | Time in days | Cost in Rs. |
| 1–2 | 6 | 1,400 | 4 | 1,900 |
| 1–3 | 8 | 2,000 | 5 | 2,800 |
| 2–3 | 4 | 1,100 | 2 | 1,500 |
| 2–4 | 3 | 800 | 2 | 1,400 |
| 3–4 | Dummy | – | – | – |

(continued)

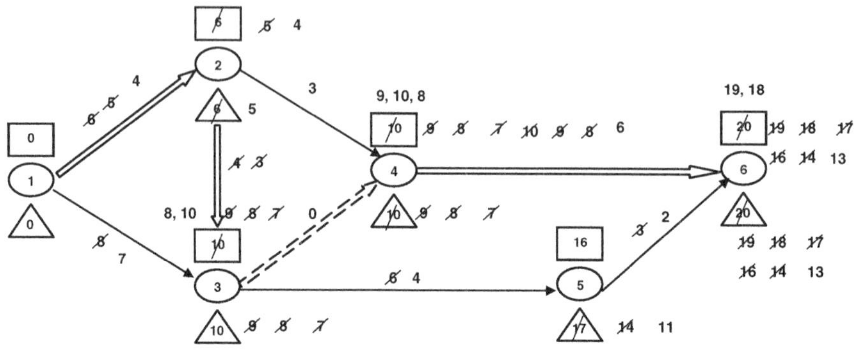

**Fig. 5.6**

(continued)

| Job | Normal time | | Crash | |
|---|---|---|---|---|
| | Time in days | Cost in Rs. | Time in days | Cost in Rs. |
| 3–5 | 6 | 900 | 3 | 1,600 |
| 4–6 | 10 | 2,500 | 6 | 3,500 |
| 5–6 | 3 | 500 | 2 | 800 |

$$\text{Total Indirect cost} = \text{Rs. } 300/\text{day} \times 20 \text{ days} = \text{Rs. } 6{,}000$$

$$\therefore \text{Normal project cost} = 9{,}200 + 6{,}000 = \text{Rs.}15{,}200$$

Now, we crash the activities and try and reduce the project duration and cost. Here, there is a limit on crashing. In below table, we have the crash time. This is the time beyond which that respective activity cannot be crashed. For example, activity 1–2 has normal time of 6 days and crash time of 4 days. This means activity 1–2 can be crashed by a maximum of 2 days and brought to 4 days, or activity 1–2 cannot be completed less than 4 days.

| Job | Normal time | | Crash | | Cost |
|---|---|---|---|---|---|
| | Time in days | Cost in Rs. | Time in days | Cost in Rs. | Slope |
| 1–2 | 6 | 1,400 | 4 | 1,900 | 250* - II |
| 1–3 | 8 | 2,000 | 5 | 2,800 | 267 |
| 2–3 | 4 | 1,100 | 2 | 1,500 | 200* - I |
| 2–4 | 3 | 800 | 2 | 1,400 | 600 |
| 3–4 | Dummy | – | – | – | -* |
| 3–5 | 6 | 900 | 3 | 1,600 | 234 |
| 4–6 | 10 | 2,500 | 6 | 3,500 | 250* - III |
| 5–6 | 3 | 500 | 2 | 800 | 300 |
| | | $\Sigma = 9{,}200$ | | | |

Next table gives the activities that are crashed and the corresponding project durations and costs. The steps followed are

| S. No. | Crash activity | Crash by days | Duration post crashing (days) | Normal cost (Rs.) | Crash cost (Rs.) | Normal + crash cost (Rs.) | Indirect cost (Rs.) | Total cost (Rs.) |
|---|---|---|---|---|---|---|---|---|
| 1 | – | – | 20 | 9,200 | – | 9,200 | 20 × 300 = 6,000 | 15,200 |
| 2 | 2–3 | 1 | 19 | 9,200 | 1 × 200 = 200 | 9,400 | 19 × 300 = 5,700 | 15,100 |
| 3 | 1–2 | 1 | 18 | 9,200 | 200 + 250 = 450 | 9,650 | 5,400 | 15,050 |
| 4 | 4–6 | 1 | 17 | 9,200 | 200 + 250 + 250 = 700 | 9,900 | 5,100 | 15,000 |
| 5 | 4–6 }<br>5–6 } | 1<br>1 | 16 | 9,200 | 1,250 | 10,450 | 4,800 | 15,250 |

Step 1: Crash activity 2–3 by 1 day, i.e., from 4 days to 3 days
Step 2: Crash activity 1–2 by 1 day
Step 3: Crash activity 4–6 by 1 day
Step 4: Crash activity 4–6 by 1 day, also activity 5–6 by 1 day. This is known as parallel crashing. All these are shown in Fig. 5.6.

From above table, we find that the optimum project duration is 17 days and optimum project cost is Rs. 15,000.

Now, if all the activities are crashed, then we have.

| S. No. | Crash activity | Crash by days | Duration post crashing (days) | Normal cost (Rs.) | Crash cost (Rs.) | Normal + Crash cost (Rs.) | Indirect cost (Rs.) | Total cost (Rs.) |
|---|---|---|---|---|---|---|---|---|
| 1 | 4–6<br>3–5 | 2<br>2 | 14 | 9,200 | 2,218 | 11,418 | 4,200 | 15,618 |
| 2 | 1–2<br>1–3 | 1<br>1 | 13 | 9,200 | 2,735 | 11,935 | 3,900 | 15,835 |

Minimum project duration is 13 days, and corresponding costs Rs. 15,835. This indicates the project cannot be completed in less than 13 days

*Note* For the crash cost column, it is always cumulative. One should not enter only the crash cost of the activity that is crashed. It should be added to the previous crash cost, figures

**Illustration 5.9**

Govil Constructions Pvt. Ltd. has submitted the following details for a particular project. It has estimated the indirect cost to be Rs. 2,000.

| Activity | Normal time (days) | Crash time (days) | Cost slope (Direct) Rs./day |
|---|---|---|---|
| 1–2 | 12 | 8 | 200 |
| 1–3 | 20 | 12 | 600 |
| 1–4 | 30 | 14 | 1,200 |
| 2–4 | 8 | 6 | 1,400 |
| 3–5 | 30 | 20 | 1,000 |
| 4–5 | 30 | 16 | 1,600 |

1. Represent the above with the help of a network diagram, and find the critical path, project duration, and normal project costs.
2. Through crashing, obtain the optimal duration and costs.

**Solution**

In this problem, we do not have direct cost. So we assume it to Rs. X. The network is shown in Fig. 5.7.

1. The critical path is 1–4–5; the normal project duration or normal project length is 60 days.

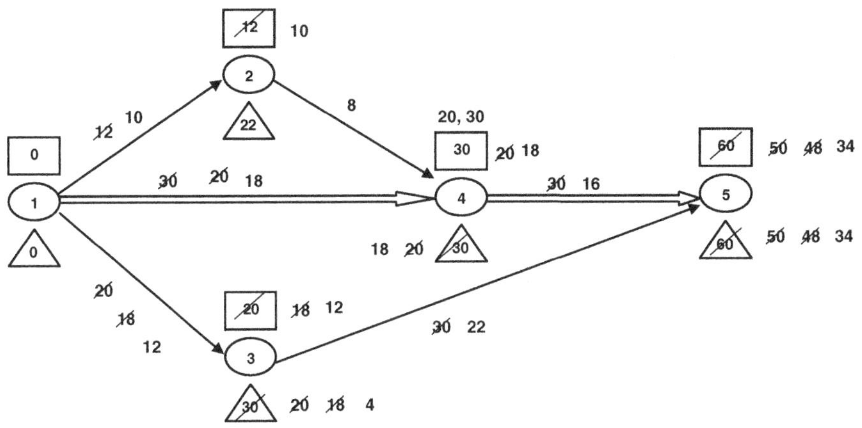


**Fig. 5.7**

$$\begin{aligned}\text{Normal projectcost} &= \text{Direct cost} + \text{Indirect cost}\\ &= X + (60 \times 2,000)\\ &= X + 1,20,000\end{aligned}$$

2. To determine the optimal duration and costs, we perform crashing as in below table.

Step 1: Crash activity 1–4 by 10 days.
Step 2: Parallel crashing—Activity 1–2 by 2 days, activity 1–3 by 2 days, and activity 1–4 by 2 days.
Step 3: Crash activity 4–5 by 14 days.
Step 4: Parallel crashing—activity 1–3 by 6 days and activity 3–5 by 8 days.

| Si. No. | Crash activity | Crash by days | Duration post crashing (days) | Normal cost (Rs.) | Crash cost (Rs.) | Normal + crash cost (Rs.) | Indirect cost (Rs.) | Total cost (Rs.) |
|---|---|---|---|---|---|---|---|---|
| 1 | – | – | 60 | X | – | X | 1,20,000 | X + 1,20,000 |
| 2 | 1–4 | 10 | 50 | X | 6,000 | X + 6,000 | 1,00,000 | X + 1,06,000 |
| 3 | 1–2 | 2 | 48 | X | 8,200 | X + 8,200 | 96,000 | X + 1,04,000 |
| | 1–3 | 2 | | | | | | |
| | 1–4 | 2 | | | | | | |
| 4 | 4–5 | 14 | 34 | X | 19,400 | X + 19,400 | 68,000 | X + 87,400 |
| 5 | 1–3 | 6 | 34 | X | 31,000 | X + 31,000 | 68,000 | X + 99,000 |
| | 3–5 | 8 | | | | | | |

From above table, we conclude that the optimal project duration is 34 days and the optimal cost or minimum cost is Rs. $X + 87{,}400$.

(*Note* If all the activities are to be crashed to the maximum extent, then crash all the activities that reduce the project duration up to the crash time that is provided, without worrying about the total project costs).

## 5.4 Flow in Networks

### *5.4.1 Maximal Spanning Tree Algorithm*

The objective in such a problem is to minimize the total distance on all the arcs of the network.

For example, laying of telephone cables in arcs.

#### 5.4.1.1 Kruskal's Method

**Illustration 5.10**

Construct the network for the following data, form the distance talk, and construct the minimum spanning tree using Kruskal's method. What is the total length?.

| Activity | Distance | Activity | Distance |
|---|---|---|---|
| 1–2 | 12 | 5–6 | 26 |
| 1–3 | 14 | 5–8 | 18 |
| 2–3 | 16 | 6–7 | 10 |
| 1–4 | 20 | 6–8 | 8 |
| 2–5 | 8 | 6–9 | 16 |
| 3–4 | 12 | 6–10 | 6 |
| 3–5 | 22 | 7–9 | 20 |
| 3–6 | 6 | 8–9 | 20 |
| 3–7 | 10 | 9–10 | 18 |
| 4–7 | 14 | | |

**Solution** From the data given, we arrange all the activities in ascending order (lowest to highest) as shown in the next table.

| Activity | Distance | Activity | Distance |
|---|---|---|---|
| 3–6 | 6 | 2–3 | 16 |
| 6–10 | 6 | 6–9 | 16 |
| 2–5 | 8 | 5–8 | 18 |
| 6–8 | 8 | 9–10 | 18 |
| 3–7 | 10 | 1–4 | 20 |

(continued)

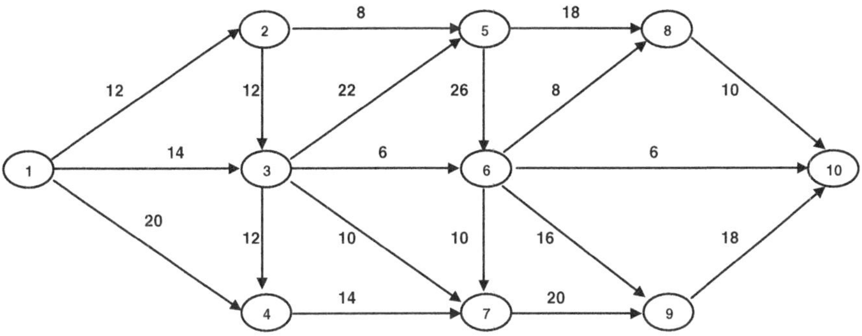


**Fig. 5.8**

(continued)

| Activity | Distance | Activity | Distance |
|---|---|---|---|
| 6–7 | 10 | 7–9 | 20 |
| 1–2 | 12 | 8–10 | 20 |
| 3–4 | 12 | 3–5 | 22 |
| 1–3 | 14 | 5–6 | 26 |
| 4–7 | 14 | | |

We draw the network as shown in Fig. 5.7.

After we draw the network, the next step is to represent only the events without any connections (Fig. 5.8).

After this, we connect the events according to ascending order. While connecting these events, it is important to remember that no closed loops are formed. Now, if we connect these events (circles) according to last table, we first connect 3–6. Then, we connect as shown in the next table.

| Priority | Activity | Distance |
|---|---|---|
| 1 | 3–6 | 6 |
| 2 | 6–10 | 6 |
| 3 | 2–5 | 8 |
| 4 | 6–8 | 8 |
| 5 | 3–7 | 10 |
| 6 | 1–2 | 12 |
| 7 | 3–4 | 12 |
| 8 | 1–3 | 14 |
| 9 | 6–9 | 16 |
| | | 92 |

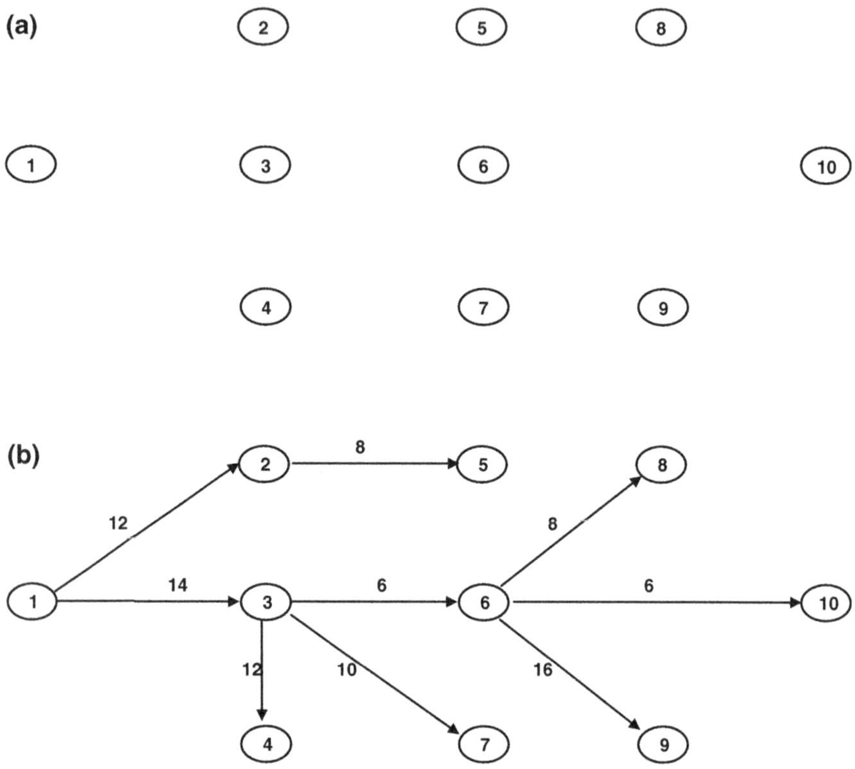


**Fig. 5.9**

The minimal spanning tree is shown in Fig. 5.9a.

In Fig. 5.9b, we find there are many open spaces. This is to ensure that no closed loop formation is taking place. For example, we cannot join 5–8, as 1–2–5–8–6–3–1 will be a closed loop. Similarly, we cannot join 2–3 as 1–2–3–1 will be a closed loop again. In the same way, it has to be ensured that no closed loop is formed. We conclude by saying the total sum of the distances is 92 or the total length is 92.

## 5.5 Review Questions

1. What is a minimum spanning tree problem? What are some of its practical applications?
2. Explain the following terms with respect to a network

   (a) Activity (b) Event.

3. What are the different types of activities and events that can be witnessed in a network representation? Explain.
4. List the rules for drawing a network.
5. Differentiate PERT and CPM.
6. Write a brief note on critical path in a project network.
7. Define the following terms:

| | | |
|---|---|---|
| (a) Total float | (b) Free float | (c) Independent float |
| (d) Head slack | (e) Tail slack | |

8. Briefly write about the demands of crashing.
9. Explain the following terms in PERT/CPM:

| | |
|---|---|
| (a) Earliest activity time | (c) Latest activity time |
| (b) Total slack | (d) Event slack |

10. Construct the project network for the following data:

| Activity | Distance |
|---|---|
| 1–2 | 10 |
| 1–3 | 16 |
| 2–4 | 12 |
| 2–5 | 8 |
| 2–6 | 8 |
| 3–7 | 10 |
| 3–8 | 6 |
| 4–9 | 2 |
| 5–9 | 6 |
| 6–10 | 10 |
| 7–10 | 8 |
| 8–11 | 18 |
| 9–12 | 4 |
| 10–12 | 8 |
| 11–13 | 2 |
| 12–13 | 14 |

11. A small project is composed of seven activities whose time estimates are listed in the time below. Activities are identified by their beginning ($i$) and ending ($j$) node numbers.

| Activity (*i–j*) | Estimated duration (Weeks) | | |
|---|---|---|---|
| | Optimistic | Pessimistic | Most likely |
| 1–2 | 1 | 1 | 7 |
| 1–3 | 1 | 4 | 7 |
| 1–4 | 2 | 2 | 8 |
| 2–5 | 1 | 1 | 1 |
| 3–5 | 2 | 5 | 14 |
| 4–6 | 2 | 5 | 8 |
| 5–6 | 1 | 6 | 15 |

(a) Draw the project network and indicate the critical path.
(b) Find the expected duration and variance of each activity.
(c) Calculate the early and late occurrence time for each node, estimated project length.
(d) What is the probability that the project will be completed?

i. At least 4 weeks earlier than expected time
ii. No more than 4 weeks later than the expected time.

12. For the data given below, find the total project duration by drawing the network diagram. List the activities that lie on the critical path.

| Activity | Description | Predecessor | Duration (in days) |
|---|---|---|---|
| 1 | Problem formulation | – | 30 |
| 2 | Literature review | – | 30 |
| 3 | Initial proposal | 1,2 | 10 |
| 4 | Industry selection | 2,3 | 15 |
| 5 | Formal research work | 2 | 20 |
| 6 | Sample size determination | 5 | 4 |
| 7 | Pilot study | 4,6 | 10 |
| 8 | Comprehensive examination | 6,7 | 3 |
| 9 | Data collection and analysis | 8,6 | 40 |
| 10 | Conclusions | 9 | 30 |
| 11 | Report preparation | 10 | 25 |
| 12 | Presenting the thesis | 11 | 4 |

13. The activities of a project are tabulated below with immediate predecessor and normal and crash cost.

| Activity | Immediate predecessor | Normal | | Crash | |
|---|---|---|---|---|---|
| | | Cost (Rs.) | Time (days) | Cost (Rs.) | Time (days) |
| A | – | 200 | 3 | 400 | 2 |
| B | – | 250 | 8 | 700 | 5 |
| C | – | 320 | 5 | 380 | 4 |
| D | A | 410 | 0 | 800 | 4 |
| E | C | 600 | 2 | 670 | 1 |
| F | B, E | 400 | 6 | 950 | 1 |
| G | B, E | 550 | 12 | 1,000 | 6 |
| H | D | 300 | 11 | 450 | 9 |

(a) Draw the network corresponding to normal time.
(b) Determine the critical path, normal duration, and the cost of the project.
(c) Suitably crash the activities so that the normal duration may be reduced by 3 days at minimum cost. Also find the project cost for this shortened duration if the indirect cost per day is Rs. 25.

14. A marketing manager wants to launch a new product. The table below shows jobs, their normal time, crash time, and cost for the project.

| Job | Normal | | Crash | |
|---|---|---|---|---|
| | Time (days) | Cost (Rs.) | Time (days) | Cost (Rs.) |
| 1–2 | 12 | 2,800 × 2 | 8 | 2,800 |
| 1–3 | 16 | 4,000 × 2 | 10 | 5,600 |
| 2–3 | 8 | 2,000 × 2 | 4 | 3,000 |
| 2–4 | 6 | 1,600 × 2 | 4 | 2,800 |
| 3–4 | Dummy | – | – | – |
| 3–5 | 12 | 1,800 × 2 | 6 | 3,200 |
| 4–6 | 20 | 5,000 × 2 | 12 | 7,000 |
| 5–6 | 6 | 1,000 × 2 | 4 | 1,600 |

Indirect cost for the project is Rs. 600 per day.

(a) Draw the project network.
(b) Determine the normal duration and cost of the project.
(c) If all the activities are crashed, what will be the project duration and corresponding costs?
(d) Find the optimum duration and minimum project cost.

15. Given the following data.

(a) Draw the network diagram.
(b) Calculate the normal project duration and cost.
(c) If all the activities are crashed to their limits, determine the corresponding time and costs.

| Activity | Immediate predecessor | Normal | | Crash | |
|---|---|---|---|---|---|
| | | Time (days) | Cost (Rs.) | Time (days) | Cost (Rs.) |
| X | – | 15 | 30,000 | 12 | 36,000 |
| Y | – | 6 | 18,000 | 6 | 18,000 |
| Z | X | 12 | 24,000 | 9 | 3,000 |
| I | X | 12 | 30,000 | 9 | 45,000 |
| J | X | 9 | 33,000 | 3 | 48,000 |
| K | Z | 3 | 21,000 | 3 | 21,000 |
| P | Y, Z | 12 | 24,000 | 6 | 36,000 |
| Q | P | 6 | 26,500 | 6 | 24,000 |
| R | J, K, Q | 9 | 22,500 | 6 | 30,000 |
| S | I | 15 | 27,000 | 9 | 36,000 |

16. For the data shown, find the minimum tree using the Kruskal's algorithm, by drawing the network.

| Activity | Distance |
|---|---|
| 1–2 | 12 |
| 4–7 | 30 |
| 1–4 | 6 |
| 5–7 | 15 |
| 1–5 | 15 |
| 5–8 | 21 |
| 2–3 | 18 |
| 6–7 | 9 |
| 2–4 | 18 |
| 6–10 | 18 |
| 3–4 | 15 |
| 7–8 | 3 |
| 3–6 | 27 |
| 7–9 | 9 |
| 4–5 | 12 |
| 8–9 | 6 |
| 4–6 | 24 |
| 9–10 | 15 |

# Chapter 6
# Game Theory

The key takeaways for the reader from this chapter are listed below:

- An insight into the world of game theory and terminologies associated with it
- Different principles of game theory
- Illustrations with solutions
- Limitations of game theory.

## 6.1 Introduction

Every business situation involves competition. In such situations, taking effective decisions is vital since it impacts the sales and consequently profit-earning potentials of the organization. Decision making can be under three situations:

(a) Deterministic
(b) Probabilistic
(c) Uncertainty.

**Deterministic Situation**: Here, the data with respect to the situation are constant. They do not vary with respect to time or any other basis.

**Probabilistic Situation**: In this case, the data with respect to the situation vary due to chance. It is possible to represent the data in some probability distribution. These probability distributions' decisions become a component of decision-making models.

**Uncertainty Situation**: As opposed to the above two, the data with respect to the situation not only vary but also do not represent any form of probabilistic distribution. Game theory is an example of uncertainty situation.

R. Srinivasan, *Strategic Business Decisions*,
DOI: 10.1007/978-81-322-1901-9_6, 

### *6.1.1 Terminologies*

**Players**: A game has two players. Companies $A$ and $B$ competing for tenders or even countries competing for trade gains are good examples of game-theoretic situations.

**Strategy**: It is the course of action taken by the player, for example, offering discount of 10 % and giving special prices.

Strategy can be pure and mixed. Suppose $m$ is the number of strategies of player **A**, $n$ is the number of strategies of player **B**, $P_i$ is the probability of selection of alternative $i$ by **A**, $i = 1, 2, 3, \ldots, m$, and $Q_j$ is the probability of selection of alternative $j$ by player **B**, $j = 1, 2, 3, \ldots, n$, then $\sum_{i=1}^{m} P_i = 1$ and $\sum_{j=1}^{n} Q_j = 1$ (sum of probabilities of selection of various alternatives for each of the players is equal to 1).

**Pure Strategy**: Suppose a player selects a particular strategy with probability equal to 1, it is referred to as pure strategy, i.e., the player is selecting that strategy alone ignoring remaining strategies. In such an event, we have

$$P_1 = 0, P_2 = 1, P_3 = 0.$$

The sum of these probabilities is equal to 1, i.e.,

$$P_1 + P_2 + P_3 = 0 + 1 + 0 = 1.$$

**Mixed Strategy**: Suppose a player follows more than one strategy, it is referred to as mixed strategy. In such a case, the probability of selection of individual strategies will be less than 1, but the sum of the probabilities is equal to 1. A sample set of the probabilities could be as follows: $Q_1 = 0.35, Q_2 = 0, Q_3 = 0.65$, and $Q_1 + Q_2 + Q_3 = 0.35 + 0 + 0.65 = 1$.

**Payoff Matrix**: Each combination of alternatives of $A$ and $B$ results in an outcome, $A_{ij}$. If $A_{ij}$ is positive, it indicates gain to $A$ and loss to $B$. If $A_{ij}$ is negative, it means loss to $A$ and gain to $B$. A payoff matrix for $A$ is as under:

Player $B$

$$\text{Player } A \quad \begin{bmatrix} a_{11} & a_{12} & \cdots & \cdots & a_{1n} \\ a_{21} & a_{22} & \cdots & \cdots & a_{2n} \\ \vdots & \vdots & \vdots & \vdots & \vdots \\ a_{m1} & a_{m2} & \cdots & \cdots & a_{mn} \end{bmatrix}$$

**Maximin Principle**: Here, gains of $A$ are maximized, i.e., it maximizes the minimum guaranteed gains of $A$. The minimum gains with respect to various alternatives of $A$ are obtained first. The maximum of these minimum gains is referred to as maximin value, and the corresponding strategy is referred to as maximin strategy.

**Minimax Principle**: Here, the maximum losses are minimized. The maximum losses of $B$, irrespective of $A$'s alternatives, are obtained first. The minimum of these maximum losses is called minimax value, and the corresponding alternative is known as minimax strategy.

**Saddle Point**: If in a game, the maximum value is equal to the minimum value, then it is said to have a saddle point. The intersecting cell corresponding to these values is referred to as the saddle point. In such a case, each player has a pure strategy.

**Value of the Game**: If there is a saddle point for the game, then the value of the cell at this point is called value of the game.

**Two-Person Zero-Sum Game**: If in a game of two players, the gain of one player is equal to the loss of the other player, then it is referred to as two-person zero-sum game. These games are also known as rectangular games, represented by a 2 × 2 matrix. In a two-person zero-sum game, there will be only two players having finite strategies to be employed against each other. Whatever may be the strategy the players use, it results in a payoff. In a two-person zero-sum game, the cell entries in $B$'s payoff matrix will be the negative of the corresponding entries in $A$'s payoff matrix.

**Illustration 6.1**

| | | Player B | | Row Minima | |
|---|---|---|---|---|---|
| | | $B_1$ | $B_2$ | ↓ | |
| Player A | $A_1$ | 18 | 4 | 4 | |
| | $A_2$ | 16 | 12 | 12 | →Maximin (Maximum of minimum profits) |
| | $A_3$ | 12 | 8 | 8 | |
| Column Maxima → | | 18 | 12 | | |
| | | | ↑ Minimax (Minimum of maximum losses) | | |

If in a game minimax = maximin, that particular point is called saddle point. The value of the game is the value of the saddle point.

A game is called a fair game if the value of the saddle point is zero. A game is called an unfair game if the value of the game is not equal to zero.

| | | Player B $B_1$ | $B_2$ | Row Minima ↓ | |
|---|---|---|---|---|---|
| Player A | $A_1$ | 18* | 4+ | 4 | |
| | $A_2$ | 16 | 12*+ | 12 | →Maximin (Maximum of minimum profits) |
| | $A_3$ | 12 | 8+ | 8 | |
| Column Maxima → | | 18 | 12 | | |
| | | | ↑ Minimax (Minimum of maximum losses) | | |

Mark the row minima with (+) and column maxima with (*). The point where (+) and (*) are marked is the saddle point, i.e., minimax = maximin

## 6.2 Games with Saddle Points

As mentioned earlier, if a game has saddle point, it is said to have pure strategy for each of its players. We can illustrate this by an example.

**Illustration 6.2**

Determine the saddle point and value for the following games. Also state whether it is a fair or an unfair game.

(I)

| | $B_1$ | $B_2$ | $B_3$ |
|---|---|---|---|
| $A_1$ | - 1 | 2 | - 2 |
| $A_2$ | 6 | 4 | - 6 |

(II)

| | $B_1$ | $B_2$ |
|---|---|---|
| $A_1$ | - 6 | 2 |
| $A_2$ | 6 | - 2 |

(III)

| | $B_1$ | $B_2$ | $B_3$ |
|---|---|---|---|
| $A_1$ | 4 | 8 | 10 |
| $A_2$ | 20 | 14 | $x$ |
| $A_3$ | 8 | $y$ | 16 |

(IV)

| | $B_1$ | $B_2$ | $B_3$ | $B_4$ | $B_5$ |
|---|---|---|---|---|---|
| $A_1$ | 6 | - 2 | 8 | 12 | 14 |
| $A_2$ | - 2 | 16 | 4 | 8 | 24 |
| $A_3$ | 32 | 16 | 12 | 28 | 24 |
| $A_4$ | 2 | 22 | -8 | 4 | 2 |

**Solution**

(I)

| | $B_1$ | $B_2$ | $B_3$ |
|---|---|---|---|
| $A_1$ | - 1 | 2 | - 2+* |
| $A_2$ | 6* | 4* | - 6+ |

The saddle point $= A_1B_3$
Value of the game $= -2$; since the value $\neq 0$, it is an unfair game. Since the value is negative, it indicates that this gives profit for player $B$ and loss for player $A$.

(II)

| | $B_1$ | $B_2$ |
|---|---|---|
| $A_1$ | - 6+ | 2* |
| $A_2$ | 6* | - 2+ |

This game does not have a saddle point as the row minima $\neq$ column maxima or minimax $\neq$ maximin.

(III)

| | $B_1$ | $B_2$ | $B_3$ |
|---|---|---|---|
| $A_1$ | 4+ | 8 | 10 |
| $A_2$ | 20* | 14*+ | $x$ |
| $A_3$ | 8+ | $y$ | 16* |

In this game, we have two variables $x$ and $y$. But for determining the saddle point, we neglect these variables and find the row minima and column maxima. We have the saddle point as $A_2B_2$, and the value of the game is 14. By neglecting $x$ and $y$, we assume that the value of $x \geq 14$ by inspecting the rows and the value of $y \leq 14$ by inspecting the columns.

Since the value of the game $\neq 0$, we can conclude that it is an unfair game.

(IV)

| | $B_1$ | $B_2$ | $B_3$ | $B_4$ | $B_5$ |
|---|---|---|---|---|---|
| $A_1$ | 6 | - 2+ | 8 | 12 | 14 |
| $A_2$ | - 2+ | 16 | 4 | 8 | 24* |
| $A_3$ | 32* | 16 | 12*+ | 28* | 24 |
| $A_4$ | 2 | 22* | -8+ | 4 | 2 |

From above table, we have saddle point $= A_3B_3$. The value of the game is 12, and it is an unfair game. Player $A$ is profiting over player $B$ here as we can see that the value of the game is positive.

## 6.3 Games with Mixed Strategies

If a game has **no** saddle point, it is said to have mixed strategies. We first solve the algorithm to determine mixed strategies followed by examples.

There can be instances where a 2 × 2, i.e., rectangular game does not have a saddle point. In such cases, we use the following formula to find the value of the game. Consider a 2 × 2 rectangular game as in Table 6.1.

**Table 6.1** Payoff matrix

| | $B_1$ | $B_2$ |
|---|---|---|
| $A_1$ | $a_{11}$ | $a_{12}$ |
| $A_2$ | $a_{21}$ | $a_{22}$ |

Let $p_1$ and $p_2$ be the probabilities with which player $A$ selects his alternatives. Let $q_1$ and $q_2$ be the probabilities with which player $B$ selects his alternatives. We have

$$p_1 = \frac{a_{22} - a_{21}}{(a_{11} + a_{22}) - (a_{21} + a_{12})} \tag{6.1}$$

and

$$p_2 = 1 - p_1 \tag{6.2}$$

Similarly,

$$q_1 = \frac{a_{22} - a_{12}}{(a_{11} + a_{22}) - (a_{21} + a_{12})} \tag{6.3}$$

and

$$q_2 = 1 - q_1 \tag{6.4}$$

The value of the game is

$$v = \frac{(a_{22} \times a_{11}) - (a_{21} \times a_{12})}{(a_{11} \times a_{22}) - (a_{21} \times a_{12})} \tag{6.5}$$

Let us consider some examples to make the concept clear.

**Illustration 6.3**
Find the value of the following games:

**Solution**
(1)

| | | Player B I | Player B II |
|---|---|---|---|
| Player A | I | 1+ | 3* |
| | II | 4* | 2+ |

(1)

| | | Player B I | Player B II |
|---|---|---|---|
| Player A | I | 1 | 3 |
| | II | 4 | 2 |

(2)

| | | Player B I | Player B II |
|---|---|---|---|
| Player A | I | - 4 | 6 |
| | II | 2 | - 3 |

We see from first table that there is no saddle point.
Therefore,

$$p_1 = \frac{2-3}{(1+2)-(4+3)} = \frac{+1}{4} \quad p_2 = 1 - \frac{1}{4} = \frac{3}{4}$$
$$q_1 = \frac{2-4}{(1+2)-(4+3)} = \frac{1}{2} \quad p_2 = 1 - \frac{1}{2} = \frac{1}{2}$$

The value of the game is $(v) = \frac{(2\times1)-(4\times3)}{(1+2)-(4+3)} = \frac{-10}{4} = \frac{5}{2}$.

Therefore, the optimal strategy for player $A = A_0$ (1/4, 3/4).
Optimal strategy for player $B = B_0$ (1/2, 1/2).
Value of the game = 5/2.
Here, $(A_0, B_0)$ are known as strategic saddle points.

(2)

| | | Player B I | Player B II |
|---|---|---|---|
| Player A | I | - 4+ | 6* |
| | II | 2* | - 3+ |

Here again, there is no saddle point, so

$$p_1 = \frac{-3-2}{(-3-4)-(2+6)} = \frac{-5}{-15} = \frac{1}{3} \quad p_2 = 1 - \frac{1}{3} = \frac{2}{3}$$
$$q_1 = \frac{-3-6}{(-3-4)-(2+6)} = \frac{-9}{-15} = \frac{3}{5} \quad p_2 = 1 - \frac{3}{5} = \frac{2}{5}$$

The value of the game is $(v) = \frac{(-3\times-4)-(2\times6)}{(-3-4)-(2+6)} = 0.$

The optimal strategy for player $A = A_0$ (1/3, 2/3).
The optimal strategy for player $B = B_0$ (3/5, 2/5).
Value of the game = 0.

**Illustration 6.4**
Two players $A$ and $B$ have a bet for tossing two coins. If both are heads, then $A$ wins Rs. 16, and if both are tails, still $A$ wins Rs. 2. If there is head first, followed by a tail, $B$ wins Rs. 6, and if the result is tail, followed by a head, $B$ wins Rs. 6. Given the choice of substituting for either player $A$ or player $B$, what would be your decision and the corresponding strategy?

**Solution**
We have the payoff matrix as in Table 6.2.

**Table 6.2** Payoff matrix between players $A$ and $B$

| | | Player B H | Player B T |
|---|---|---|---|
| Player A | H | 16 | -6 |
| | T | -6 | 2 |

We notice that there is no saddle point. So we compute $p_1$, $p_2$, $q_1$, and $q_2$.

$$p_1 = \frac{2-(-6)}{(2+16)-(-6-6)} = \frac{8}{30} = \frac{4}{15} \quad p_2 = 1 - \frac{4}{15} = \frac{11}{15}$$
$$q_1 = \frac{2-(-6)}{(2+16)-(-6-6)} = \frac{8}{30} = \frac{4}{15} \quad q_2 = 1 - \frac{4}{15} = \frac{11}{15}$$

The value of the game is $(v) = \frac{(2\times16)-(-6\times-6)}{(2+16)-(-6-6)} = \frac{-4}{30} = \frac{-2}{15}$

As the value of the game is negative, it indicates that player $B$ makes profits or wins. So the choice would be to choose player $B$.

The optimal strategy for player $A$ = $A_0$ (4/15, 11/15).
The optimal strategy for player $B$ = $B_0$ (4/15, 11/15).
Value of the game = −2/15.

## 6.4 Dominance Principle

It is possible, in some games, to reduce the size of the payoff matrix by eliminating redundant rows or columns. When a game has such rows or columns, they are dominated by some rows or columns, respectively, i.e., for a payoff matrix of larger size, the rule of dominance can be used to reduce its size by carefully eliminating certain row and/or columns prior to final analysis to determine the optimal strategy for each player. This is referred to as dominance property.

Generally, the following rules are used to reduce the size of the payoff matrix.

### *6.4.1 Dominance Property for Rows*

If all the elements in the $i$th row of the payoff matrix are less than or equal to the corresponding elements in the $j$th row in the payoff matrix, then player $A$ will never choose the $i$th strategy, or the $i$th strategy is dominated by the $j$th strategy.

### *6.4.2 Dominance Property for Columns*

If all the elements in the $p$th column of the payoff matrix are greater than or equal to the corresponding elements in the $q$th column, then player $B$ will never choose the $p$th strategy, or the $p$th strategy is dominated by the $q$th strategy.

We illustrate the above concept with some examples.

**Illustration 6.5**
Solve the following game through dominance principle:

| | | Player $B$ | | | | |
|---|---|---|---|---|---|---|
| | | 1 | 2 | 3 | 4 | 5 |
| Player $A$ | 1 | 6 | 10 | 8 | 18 | 12 |
| | 2 | 10 | 12 | 6 | 14 | 16 |
| | 3 | 16 | 14 | 18 | 16 | 14 |
| | 4 | 8 | 4 | 16 | 10 | 6 |

**Solution**

By the dominance principle, we eliminate as many rows and/or columns through comparison. We solve the above problem by adopting the following steps:

Step 1 Compare row 3 with row 4. All the elements in row 3 are greater than the corresponding elements in row 4. Therefore, we delete row 4 (∵ the objective of player *A* is to maximize profits).

Step 2 Compare column 2 with column 4. All the elements in column 4 are greater than the corresponding elements in column 2. Therefore, we delete column 4 (∵ the objective of player *B* is to minimize profits).

Step 3 Compare column 3 with column 5; all elements in 5 are greater than the corresponding elements in 3. Therefore, delete 5.

Step 4 Compare row 1 with row 2; this results in deleting row 1.

Step 5 Compare row 2 with row 3 and delete row 2.

Step 6 Compare column 1 and column 3 and delete column 3.

Step 7 Compare column 1 and column 2 and delete column 1.

These steps are shown in Table 6.3.

So we are left with 14.

The optimal strategy for player *A* is $A_3$.

The optimal strategy for player *B* is $B_2$.

The value of the game is $(v) = 14$.

**Table 6.3** The payoff matrix for player A

| | | | | | | Step 2 ↓ | Step 3 ↓ | |
|---|---|---|---|---|---|---|---|---|
| | | | 1 | 2 | 3 | 4 | 5 | |
| | Step 4 → | 1 | 6 | 10 | 8 | 18 | 12 | |
| | Step 5 → | 2 | 10 | 12 | 6 | 14 | 16 | |
| Player *A* | Step 7 | 3 | 16 | 14 | 18 | 16 | 14 | |
| | Step 6 | 4 | 8 | 4 | 16 | 10 | 6 | ← Step 1 |

**Illustration 6.6**

Given the payoff matrix for player *A*, solve the same by dominance principle. Obtain the optimum strategies for both players and value of the game.

Player *B*

| Player *A* | | $B_1$ | $B_2$ | $B_3$ | $B_4$ | $B_5$ |
|---|---|---|---|---|---|---|
| | $A_1$ | 2 | 4 | 3 | 8 | 4 |
| | $A_2$ | 5 | 6 | 3 | 7 | 4 |
| | $A_3$ | 6 | 7 | 9 | 8 | 7 |
| | $A_4$ | 4 | 2 | 8 | 4 | 3 |

**Solution**

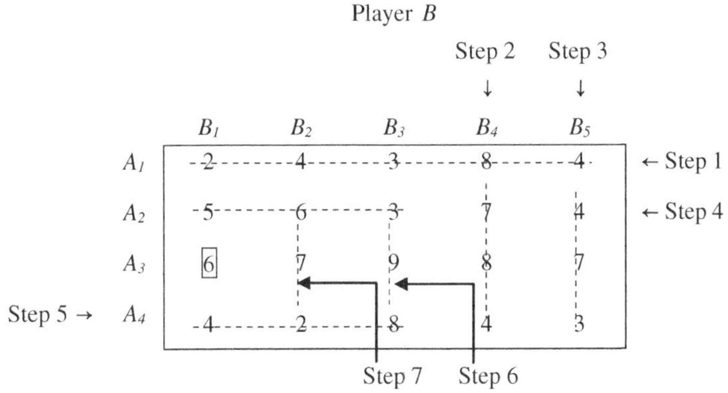


Step 1 Compare row 1 with row 3 and delete row 1
Step 2 Compare column 1 with column 4 and delete column 4
Step 3 Compare column 2 with column 5 and delete column 5
Step 4 Compare row 2 with row 3 and delete row 2
Step 5 Compare row 3 with row 4 and delete row 4
Step 6 Compare column 2 with column 3 and delete column 3
Step 7 Compare column 1 with column 2 and delete column 2.

We are left with 6. Hence,
The optimal strategy for player *A* is $A_3$.
The optimal strategy for player *B* is $B_1$.
The value of the game is $(v) = 6$.

**Illustration 6.7**
Use the relation of dominance to solve the following game, given the payoff matrix:

Player B

| Player A | $B_1$ | $B_2$ | $B_3$ | $B_4$ | $B_5$ | $B_6$ |
|---|---|---|---|---|---|---|
| $A_1$ | 0 | 0 | 0 | 0 | 0 | 0 |
| $A_2$ | 8 | 4 | 0 | 4 | 2 | 2 |
| $A_3$ | 8 | 6 | 2 | 6 | 4 | 4 |
| $A_4$ | 8 | 6 | 14 | - 10 | 2 | 4 |
| $A_5$ | 8 | 6 | 8 | - 2 | 4 | 4 |
| $A_6$ | 8 | 6 | 6 | - 4 | 4 | 4 |

**Solution**

We apply the usual dominance principle rule to solve the given problem. The steps are given below:

Player B

| Player A | $B_1$ | $B_2$ | $B_3$ | $B_4$ | $B_5$ | $B_6$ |
|---|---|---|---|---|---|---|
| $A_1$ | 0 | 0 | 0 | 0 | 0 | 0 |
| $A_2$ | 8 | 4 | 0 | 4 | 2 | 2 |
| $A_3$ | 8 | 6 | 2 | 6 | 4 | 4 |
| $A_4$ | 8 | 6 | 14 | - 10 | 2 | 4 |
| $A_5$ | 8 | 6 | 8 | - 2 | 4 | 4 |
| $A_6$ | 8 | 6 | 6 | - 4 | 4 | 4 |

Step 1 Compare row 1 with row 2 and delete row 1
Step 2 Compare row 2 with row 3 and delete row 2
Step 3 Compare column 1 with column 2 and delete column 1
Step 4 Compare column 2 with column 5 and delete column 2
Step 5 Compare row 5 with row 6 and delete row 6.

In the rest of the elements, we cannot delete any row or column. Hence we rewrite the left-out elements as in below table.

| | $B_3$ | $B_4$ | $B_5$ | $(B_3 + B_4)/2$ |
|---|---|---|---|---|
| $A_3$ | 2 | 6 | 4 | 4 |
| $A_4$ | 14 | -10 | 2 | 2 |
| $A_5$ | 8 | - 2 | 4 | 3 |

Since we cannot delete any row or column, we take the average of two columns $B_3$ and $B_4$ and write down the values on the right of the table. Compare these values with the corresponding values of other columns and delete the column with higher values. In Table 6.22, we find column 3 or $B_5$ has higher values, and so we delete it.

*Note* One thing to remember is not to delete the average values itself on comparing or for any other reason. It should be left untouched.

We write down the remaining elements ((2 × 3) matrix) in below table.

| | $B_3$ | $B_4$ |
|---|---|---|
| $A_3$ | 2 | 6 |
| $A_4$ | 14 | -10 |
| $A_5$ | 8 | - 2 |
| $(A_3 + A_4)/2$ → | 8 | -2 |

Now, we take the average of two rows $A_3$ and $A_4$ and compare these values with the corresponding values of other columns and delete row 3, i.e., $A_5$.

By doing this, we reduce the (2 × 3) game to a regular (2 × 2) game as shown in above table.

| | $B_3$ | $B_4$ |
|---|---|---|
| $A_3$ | 2 | 6 |
| $A_4$ | 14 | -10 |

Now, we compute $p_1$, $p_2$, $q_1$, and $q_2$ and value of the game.

$$p_1 = \frac{-10-14}{-8-20} = \frac{-24}{-28} = \frac{6}{7} \quad p_2 = 1 - \frac{6}{7} = \frac{1}{7}$$

$$q_1 = \frac{-10-6}{-8-20} = \frac{-16}{-28} = \frac{4}{7} \quad q_2 = 1 - \frac{4}{7} = \frac{3}{7}$$

The value of the game is $(v) = \frac{(2\times-10)-(14\times6)}{-8-20} = \frac{-104}{-28} = \frac{26}{7}$.

## Conclusions

The optimal strategy for player $A$ is $A_0$ (6/7, 1/7)
The optimal strategy for player $B$ is $B_0$ (4/7, 3/7)
The value of the game is $(v) = 26/7$.

## Illustration 6.8

Solve the following 3 × 5 game using dominance principle:

| | | Player $B$ | | | | |
|---|---|---|---|---|---|---|
| | | $B_1$ | $B_2$ | $B_3$ | $B_4$ | $B_5$ |
| | $A_1$ | 6 | 15 | 30 | 21 | 6 |
| Player $A$ | $A_2$ | 3 | 3 | 6 | 6 | 4 |
| | $A_3$ | 12 | 12 | 24 | 36 | 3 |

**Solution**

The solution is given in below table.

| | | Player $B$ | | | | | |
|---|---|---|---|---|---|---|---|
| | | $B_1$ | $B_2$ | $B_3$ | $B_4$ | $B_5$ | |
| Step 5 | $A_1$ | 6 | 15 | 30 | 21 | 6 | |
| Player $A$ | $A_2$ | 3 | 3 | 6 | 6 | 4 | ← Step 4 |
| | $A_3$ | 12 | 12 | 24 | 36 | 3 | ← Step 6 |
| | | | ↑ Step 1 | ↑ Step 3 | ↑ Step 2 | | |

Step 1 Compare column 1 with column 2 and delete column 2
Step 2 Compare column 1 with column 4 and delete column 4
Step 3 Compare column 1 with column 3 and delete column 3
Step 4 Compare row 1 with row 2 and delete row 2
Step 5 Compare column 1 with column 5 and delete column 1
Step 6 Compare row 1 with row 3 and delete row 3.

We are left with 6.

**Conclusions**

The optimal strategy for player $A$ is $A_1$.
The optimal strategy for player $B$ is $B_5$.
The value of the game is $(v) = 6$.

## 6.5 Graphical Method for $2 \times n$ or $m \times 2$ Games

**$2 \times n$ or $m \times 2$** are **games** specialized nature of games. The payoff matrix for a $2 \times n$ game will contain 2 rows and $n$ columns and that of $m \times 2$ game $m$ rows and 2 columns. If there is no saddle point, then the games can be solved using graphical method and reduced to a $2 \times 2$ rectangular game and then solved by mixed strategy principle.

## 6.5.1 Graphical Solution for 2 × n Games

**Illustration 6.9**

Solve the following game graphically:

| | | Player *B* | | |
|---|---|---|---|---|
| | | $B_1$ | $B_2$ | $B_3$ |
| Player *A* | $A_1$ | 2 | 6 | 22 |
| | $A_2$ | 16 | 10 | 4 |

**Solution**

First, we plot the graph with the point values given in above table.

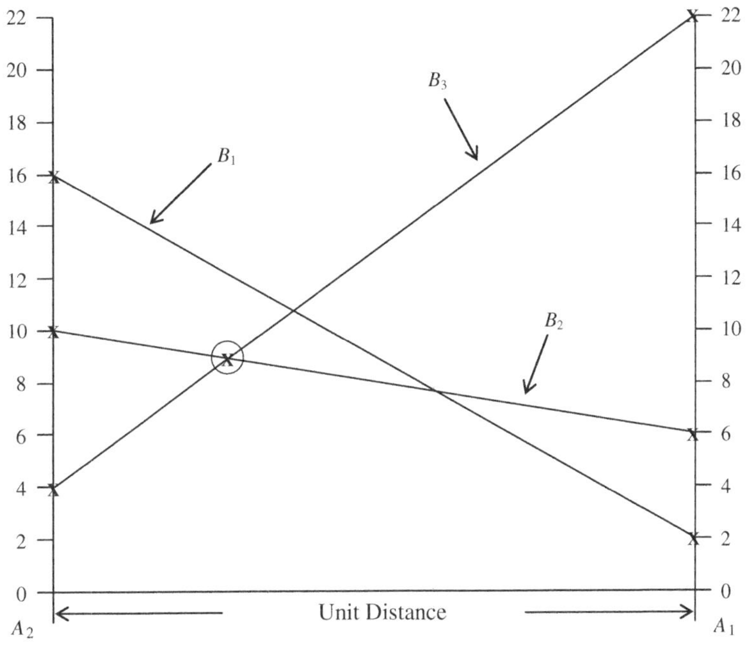


**Fig. 6.1** Graphical solution for 2 × *n* games

In Fig. 6.1, we have 3 lines, $B_1$, $B_2$, and $B_3$. Since this is a 2 × *n* game, we have to identify the maximin point (lower most boundary and upper most point). Then, we omit that line which is not involved in the maximin point. Here, $B_1$ is not involved in the maximum point, so omit $B_1$.

Now, we can form a 2 × 2 game with the values shown in next table.

| | | Player *B* | |
|---|---|---|---|
| | | $B_2$ | $B_3$ |
| Player *A* | $A_1$ | 6 | 22 |
| | $A_2$ | 10 | 4 |

For this, we use the mixed strategy principle to compute the optimal strategy and value of the game.

We have

$$p_1 = \frac{4-10}{10-32} = \frac{-6}{-22} = \frac{3}{11} \quad p_2 = 1 - \frac{3}{11} = \frac{8}{11}$$
$$q_1 = \frac{4-22}{10-32} = \frac{-18}{-22} = \frac{9}{11} \quad q_2 = 1 - \frac{9}{11} = \frac{2}{11}$$

The value of the game is $(v) = \frac{24-220}{10-32} = \frac{-196}{-22} = \frac{98}{11}$.

**Conclusions**

The optimal strategy for player *A* is $A_0$ (3/11, 8/11).
The optimal strategy for player *B* is $B_0$ (9/11, 2/11).
The value of the game is $(v) = 98/11$.

**Illustration 6.10**

Solve the following (2 × 4) game:

| | | Player *B* | | | |
|---|---|---|---|---|---|
| | | $B_1$ | $B_2$ | $B_3$ | $B_4$ |
| Player *A* | $A_1$ | 2 | 2 | 3 | -1 |
| | $A_2$ | 4 | 3 | 2 | 6 |

**Solution**

We plot the graph from the given values in the table above (Fig. 6.2).

Here, $B_2$ and $B_3$ are involved in the uppermost point of the lowermost boundary. So we get a (2 × 2) rectangular game as are involved in the uppermost point of

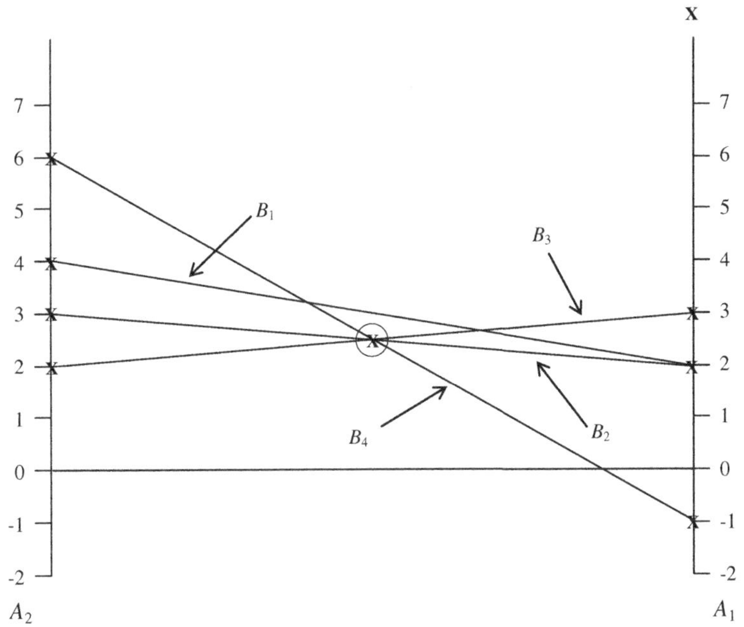


**Fig. 6.2** Graphical solution for 2 × 4 games where $B_2$ and $B_3$ are involved in the uppermost point of the lowermost boundary

lowermost

| | | Player *B* | |
|---|---|---|---|
| | | $B_2$ | $B_3$ |
| Player *A* | $A_1$ | 2 | 3 |
| | $A_2$ | 3 | 2 |

From the mixed strategy principle

$$p_1 = \frac{2-3}{4-6} = \frac{-1}{-2} = \frac{1}{2} \quad p_2 = 1 - \frac{1}{2} = \frac{1}{2}$$
$$q_1 = \frac{2-3}{4-6} = \frac{-1}{-2} = \frac{1}{2} \quad q_2 = 1 - \frac{1}{2} = \frac{1}{2}$$

The value of the game is $(v) = \frac{4-9}{4-6} = \frac{5}{2}$.

**Conclusions**

The optimal strategy for player *A* is $A_0$ (1/2, 1/2).
The optimal strategy for player *B* is $B_0$ (1/2. 1/2).
The value of thc game is $(v) = 5/2$.

## 6.6 Graphical Solution for $m \times 2$ Games

**Illustration 6.11**
Solve the following game by graphical method, given the payoff matrix in the following table:

| | | Player $B$ $B_1$ | $B_2$ |
|---|---|---|---|
| Player $A$ | $A_1$ | 4 | 8 |
| | $A_2$ | 4 | 6 |
| | $A_3$ | 6 | 4 |
| | $A_4$ | - 4 | 12 |

**Solution**
The graph is plotted for the given values as shown in Fig. 6.3.

In case of $m \times 2$ games, we have to identify the minimax point in order to reduce the given game to $2 \times 2$ game. The minimax point (uppermost boundary and lowermost point) is shown in Fig. 6.3.

Out of the four lines, $A_2$ and $A_4$ are involved in the minimax point. So we omit $B_3$ and $B_4$, and we get the $2 \times 2$ game as in Table 6.4.

**Table 6.4** Graphical method

| | | Player $B$ $B_1$ | $B_2$ |
|---|---|---|---|
| Player $A$ | $A_1$ | 4 | 8 |
| | $A_2$ | 6 | 4 |

Now, we adopt the mixed strategy rule to compute the optimal strategies and the value of the game. We have

$$p_1 = \frac{4-6}{8-14} = \frac{-2}{-6} = \frac{1}{3} \quad p_2 = 1 - \frac{1}{3} = \frac{2}{3}$$

$$q_1 = \frac{4-8}{8-14} = \frac{-4}{-6} = \frac{2}{3} \quad q_2 = 1 - \frac{2}{3} = \frac{1}{3}$$

The value of the game is $(v) = \frac{16-48}{8-14} = \frac{-32}{-6} = \frac{16}{3}$.

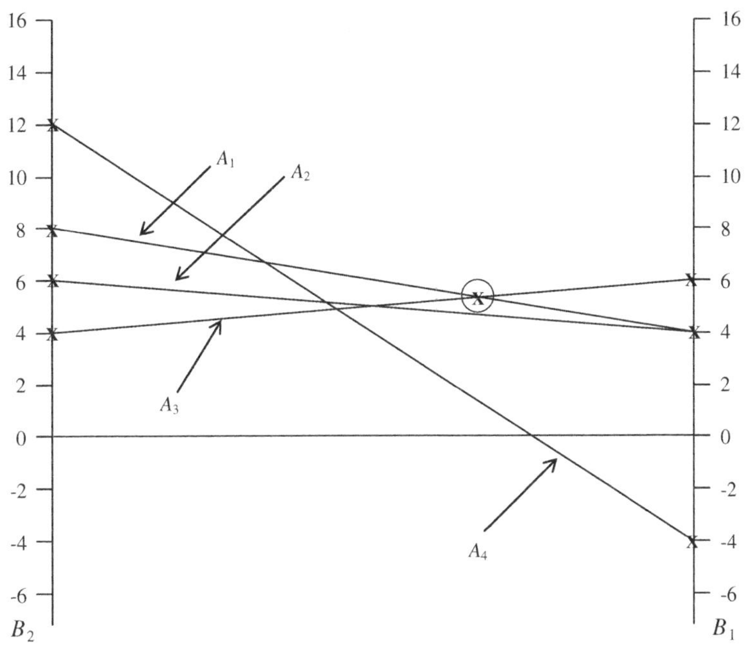


**Fig. 6.3** Graphical solution for $m \times 2$ games

## Conclusions

The optimal strategy for player $A$ is $A_0$ (1/3, 2/3).
The optimal strategy for player $B$ is $B_0$ (2/3. 1/3).
The value of the game is $(v) = 16/3$.

## Illustration 6.12

Below table shows a payoff matrix of player $A$. Solve it by graphical method.

| | | Player $B$ | |
|---|---|---|---|
| | | $B_1$ | $B_2$ |
| Player $A$ | $A_1$ | 2 | 6 |
| | $A_2$ | 6 | 2 |
| | $A_3$ | 10 | - 2 |
| | $A_4$ | 12 | - 12 |

## Solution

Figure 6.4 shows the graph for the given values.

We have $A_1$ and $A_2$ to form the lower most point in the uppermost boundary. So the rectangular matrix can be written as shown in Table 6.5.

**Table 6.5** Rectangular matrix

| | | Player *B* | |
|---|---|---|---|
| | | $B_1$ | $B_2$ |
| Player *A* | $A_1$ | 2 | 6 |
| | $A_2$ | 10 | - 2 |

Using the mixed strategy principle, we have

$$p_1 = \frac{-2-10}{0-16} = \frac{-12}{-16} = \frac{3}{4} \quad p_2 = 1 - \frac{3}{4} = \frac{1}{4}$$
$$q_1 = \frac{-2-6}{0-16} = \frac{-8}{-16} = \frac{1}{2} \quad q_2 = 1 - \frac{1}{2} = \frac{1}{2}$$

The value of the game is $(v) = \frac{-4-60}{0-16} = \frac{-64}{-16} = 4$.

**Conclusions**

The optimal strategy for player *A* is $A_0$ (3/4, 1/4).
The optimal strategy for player *B* is $B_0$ (1/2, 1/2).
The value of the game is $(v) = 4$.

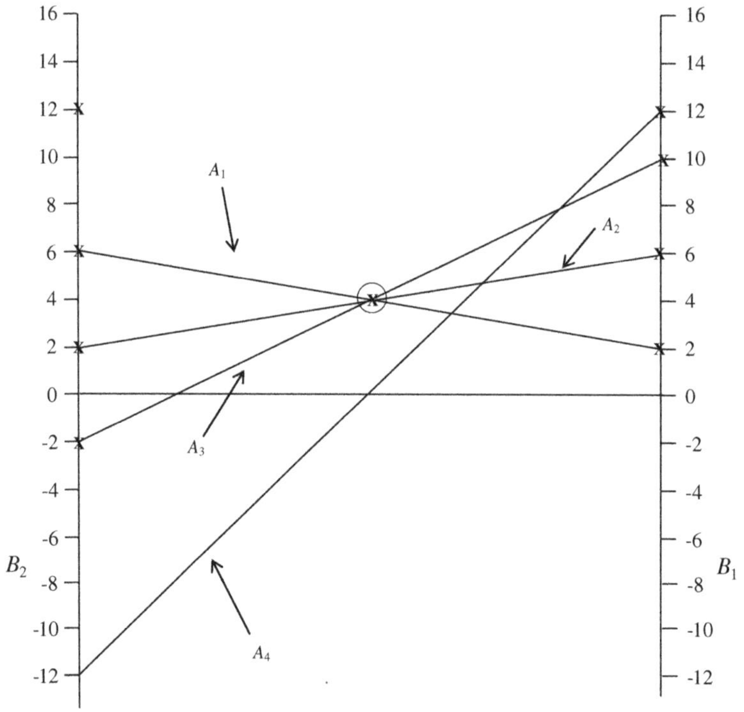


**Fig. 6.4** Graphical method solution for payoff matrix of player *A*

## 6.7 Limitations of Game Theory

Apart from many advantages of game theory, there are a few disadvantages. They are highlighted below. It is important for the user of game theory to remember this while applying this concept to real-world problems.

1. It is unrealistic to assume that the players have knowledge of their own and other's payoffs. It can at best be a guess.
2. When the number of players increases, gaming strategies become difficult and complex. Oligopolistic market could be an example where game theory is not of much help.
3. Assumptions of maximin and minimax that players are more averse and have complete knowledge of the strategies are unlikely to be true.
4. When in an oligopolistic situation, the players allow each other to share the secrets of business in order to work out collusion, mixed strategies are not very useful.

## 6.8 Advances in Game Theory

As we know by now, the decision makers in a game are usually referred to as players or agents. Game theory provides for general quantitative methodologies for analyzing situations where two or more players decide moves to affect each other's welfare. We first introduce a class of games called strategic form games which are most common among all types of games.

Strategic form games: Strategies imply actions of players or more specifically pure strategies. The set $P$ is a collection of all strategy profiles of players denoted by the Cartesian product $P_1 \times P_2 \times \cdots \times P_n$. The utility of a player depends not only on his own strategy but also on the strategies of the rest of the players. Every strategy profile induces an outcome in the game. If the strategy profiles are finite, then we call the game as finite. A strategic form game captures each player's decision problem of choosing a strategy. A strategic form game can be represented as follows: $N, P_i(i \epsilon N), u_i(i \epsilon N)$ where $N$ is the total number of players and $\mu_i$ is the utility or payoff function.

## 6.9 Notions in Game Theory

The following notions are fundamental to game theory other than utilities. Let us discuss these notions and a few related issues.

**Intelligence**: Intelligence of players is a key notion in game theory. It connotes that each participating player knows everything about the game and the game designer knows the players are competent enough to draw inferences about the game that the designer can make. This type of action is called strategic, which means an intelligent player would fully take into account his knowledge or expectation of behavior of other players in determining what his best response would be. It is assumed that each player has the required resources and computations to determine the best strategy.

**Common Knowledge**: An implication of common knowledge is intelligence. Robert Aumann defines common knowledge as follows: A fact is common knowledge among the players if every player knows it, every player knows that every player knows it, and so on. That is every statement of the form "every player knows that every player knows that…every player knows it" is true and infinitum. If it happens that a fact is known to all the players without the requirement of all players knowing that all players know it, etc., then such a fact is mutual knowledge. Game-theoretic situations often require this to be true. If a player has private information, then that is not common knowledge.

## *6.9.1 Example for Illustrating Common Knowledge*

**Illustration 6.13**

Assume that there are five rational and professional software developers (reportees) P, Q, R, S, and T and intelligent supervisors p, q, r, s, and t. There are five separate modules being developed, which need to be integrated later; the supervisors get to meet everyday to get a status update and performance of their reportees. Every day when the supervisors meet, the protocol is the following. If the supervisor thinks his reportee is progressing satisfactorily, he will praise his virtues. On the other hand, if the supervisor thinks his reportee's performance is not up to the mark, he will criticize him or her. All the supervisors follow this protocol.

The fact is that none of the developers are progressing satisfactorily, but they are disguising their performance from their supervisors. However, if a supervisor finds that the reportee of another supervisor is not progressing satisfactorily, he will immediately report it to all supervisors except the reportee's supervisor. For example, if p finds Q short of performance, he will report to r, s, and t but not to q. This protocol is also known to all supervisors. Since none of the reportees are progressing satisfactorily, the fact that a reportee is not progressing satisfactorily is common knowledge among all the supervisors except the reportee's supervisor.

Since each supervisor does not know that his reportee is progressing badly, it turns out that he always provides a positive recommendation every day. On a fine day, the senior manager meets all supervisors and makes the following statement: "one of the reportees is not progressing satisfactorily."

Note that this becomes common knowledge among all the supervisors.

## 6.10 Classification of Games

There are many classes of games; we provide a listing of some of the well-known games here.

### *6.10.1 Cooperative and Non-cooperative Games*

A game is cooperative if players agree to cooperate; that is, agreements, promises, or threats are enforceable among players. We can see joint group of actions in cooperative games. If the commitments, promises, or threats are not enforceable, then the game is non-cooperative. Here, each player chooses his action individually. In a strategic form game, a player chooses his action only once and all players exercise their decisions simultaneously. If we can associate a value with some subset of players, then this kind of game becomes a coalition game. This is appropriate for cooperative games.

### *6.10.2 Games with Complete and Incomplete Information*

If every aspect of the game is common knowledge, then it is said to be a game with complete information. On the other hand, at the point when players begin to plan their moves, if some players have some private information that other players do not know, then it is said to be a game with incomplete information.

### *6.10.3 Games with Perfect and Imperfect Information*

Before beginning the game if a player knows the history, i.e., his own and other players past moves, then the game is said to be perfect, else it is said to be a game with imperfect information.

## 6.11 The Notion of Strategy

In game theory, the notion of strategy is very important. A strategy can be described as a complete plan or action set which is indicative of what a player will do at each step where the player is called upon to play. A strategy specifies the action a player chooses at every point during the game.

### 6.11.1 Examples of Strategic Games

**Illustration 6.14**
Prisoner's Dilemma

One of the widely studied problems in game theory is the prisoner's dilemma problem with many interesting interpretations in a wide variety of situations. Two criminals are arrested for committing a crime and are lodged in separate prisons where they are interrogated by the prison superintendent. The superintendent informs the prisoners in private that if he confesses, he would get a shorter jail sentence of 1 year, while the other would be sentenced for 10 years. If both prisoners confess, they both would get a 5-year sentence in jail. If neither confesses, they would be sentenced to 2 years. The superintendent also informs each prisoner what has been told to the other prisoner. Thus, the utility or payoff matrix is common knowledge to both the prisoners as given in

| 1 | 2 | |
|---|---|---|
| | NC | C |
| NC | -2, -2 | -10, -1 |
| C | -1, -10 | -5, -5 |

Here, each player (prisoner) adopts the best response strategy in response to the other player's best response. One can observe that $C$ would be the best strategy, regardless of what the other player plays:

$$u_1(C,C) > u_1(NC,C); u_2(C,C) > u_2(NC,C); u_2(NC,C) > u_2(NC,NC); u_1(C,NC) > u_1(NC,NC)$$

Thus, ($C$, $C$) is a natural prediction for the game. The outcome ($NC$, $NC$) is the best outcome jointly for the players, provided that each player has private information that the other player also adopts NC. Prisoner's dilemma is a classic example of a rational and intelligent game that does not produce socially optimal outcomes. Each prisoner has a negative effect on the other. If a prisoner chooses to move away from ($NC$, $NC$), he would reduce his jail sentence by 1 year while increasing the other player's sentence by 8 years.

**Illustration 6.15**
Cold war (competition) between countries

Two countries X and Y are competing with each other on their spending in defense and health care. The payoff matrix is given below:

| X | Y | |
|---|---|---|
| | **Healthcare** | **Defense** |
| **Healthcare** | 5,-5 | -5, 10 |
| **Defense** | 10, -5 | 0,0 |

Here, we can observe that defense would be the best response irrespective of the other country's strategy. Also, a country can choose health care only if the other country chooses health care too. In this game, (health care, health care) is termed as socially optimal outcome, while (defense, defense) is the predicted outcome. Socially optimal outcomes are also called Pareto efficient outcomes, which means if players cooperate, resulting outcomes would be socially optimal.

## 6.12 Extensive Form Games

This form of game was proposed by Von Neumann and Morgenstern and captures complete sequential play of the game. It captures who moves when; what players know before playing at every stage; what actions result in what outcomes; what utilities follow the outcome; and more importantly what actions each player may play. Generally, extensive form games are depicted graphically using trees (usually referred as game trees).

**Illustration 6.16**

Matching pennies

In this game, there are two players 1 and 2 who have a coin each. Both spin their coins and are able to see the outcome (heads or tails). If both show heads or tails, player 1 is the winner who gets \$1 from player 2. If either of them is heads or tails, then player 1 pays \$1 to player 2. The tree below shows the scenario when player 1 makes the first move. Generally, there could be three types of nodes in a tree representation: root node (first move node); decision nodes (internal nodes or subsequent decision nodes); and terminal nodes. The sequence of events is captured by a path of connections from the root to the terminal node. Each decision node is labeled with the player who makes a decision at that node. Nodes represent not only the current position in the game but also the path toward that point. The terminal nodes carry the labels of the payoffs of players.

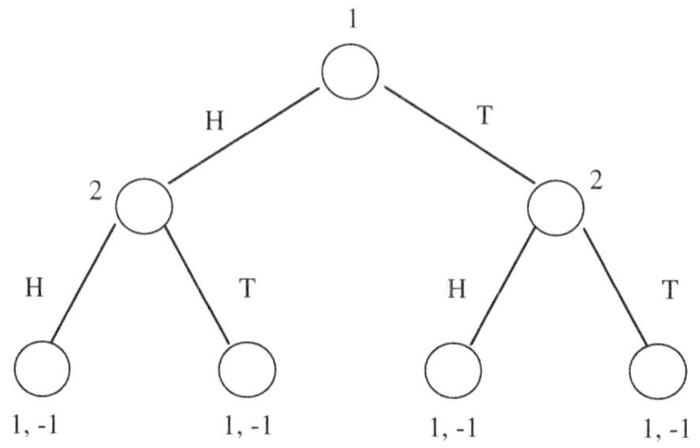


**Fig. 6.5** Matching pennies when player 1 makes the first move

The other version of the game is when players are not able to observe the outcomes. Depending on whether player 1 or player 2 makes the first move, we can draw a tree as above. Figure 6.6 shows a scenario where player 2 makes the first move.

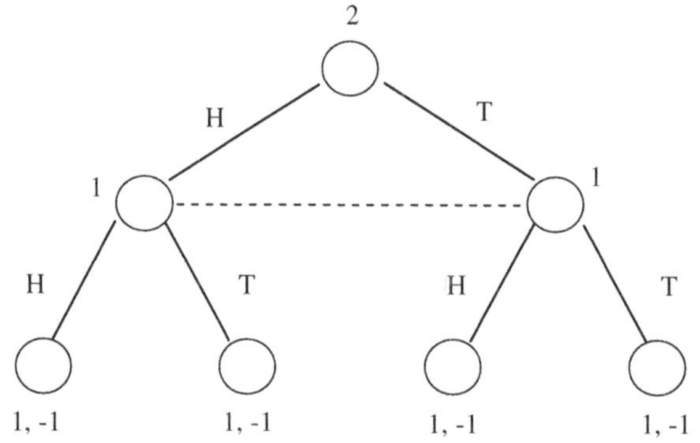


**Fig. 6.6** Matching pennies when outcomes are not observed and player 2 plays first

The tree in Fig. 6.6 is similar to that in Fig. 6.5 except that we can see a dotted line. This dotted line represents the information sets. When the game reaches to either of those nodes, the player who has to make a move does not know his information set as he has not observed the outcome of the past move. Thus, information sets of a player describe a collection of all possible distinguishable circumstances in which the player is called upon to move. The corresponding player, in every node within a given information set, must have the same set of possible actions.

The third version of the matching pennies game is when players spin their coins simultaneously. Here too, no player would get a chance to observe the outcome of the move of the other player. Both trees depicted in Figs. 6.5 and 6.6 become a valid representation of the game as order of the game is not relevant here.

## 6.13 Games with Perfect and Imperfect Information

If all the information sets are singletons, then the game is said to be with perfect information. A direct implication is that players would be able to observe all the outcomes and know the complete history thus far in the game. If any of the information contains more than one element, then the game is said to be with imperfect information. Figure 6.5 represents a game with perfect information, and Fig. 6.6 represents a game with imperfect information.

An extensive form game $G$ with perfect information consists of $N$ players, their action sets $A_i$, set of terminal histories $H$, and utilities of the players corresponding to each terminal history $U_i$, and $P{:}S_h \rightarrow N$ is a mapping of each sub-history to a player (including empty histories). It is represented as follows:

$$G = (N, A_i, H, P, U_i)$$

For Fig. 6.5, when we apply the above definition, we get the following:

$$\begin{aligned} N &= \{1, 2\} \\ A_i &= \{\mathrm{H}, \mathrm{T}\} \forall i \\ H &= \{\mathrm{HH}, \mathrm{HT}, \mathrm{TH}, \mathrm{TT}\} \\ S_h &= \{\mathrm{H}, \mathrm{T}, \mathrm{E}\} \text{ where } E \text{ is the empty history} \\ P(E) &= P(\mathrm{T}) = 2; P(E) = 1 \end{aligned}$$

$U_i$s are shown in below table.

| | Player 2 | |
|---|---|---|
| **Player 1** | **H** | **T** |
| **H** | 1, -1 | -1, 1 |
| **T** | -1, 1 | 1, -1 |

## 6.14 How to Define Strategy?

The notion of strategy is important in game theory. A strategy can be defined as a complete action plan that will specify what each player will do at each of the information sets where the player is called upon to play. Specifically, a strategy set $S_i$ is a mapping of information sets $I_i \rightarrow A_i$ such that $S_i(K) \in C(K) \; \forall \; K \in I_i$. $K$ is the information set drawn from $I_i$, and $M(K) \subset A_i$ is the set of all possible actions to player $i$ in the information set $K$.

**The notion of strategy with respect to** Fig. 6.6 **where player 1 moves first**

Here, $I_1 = \{E\}$ and $I_2 = \{\text{H, T}\}$. Both the players have two strategies as shown below.

$$S_{11} : \{E\} \rightarrow \text{H}$$
$$S_{12} : \{E\} \rightarrow \text{T}$$
$$S_{21} : \{\text{H}, \text{T}\} \rightarrow \text{H}$$
$$S_{22} : \{\text{H}, \text{T}\} \rightarrow \text{T}$$

The payoff matrix can be formulated from the above derivations as illustrated above.

## 6.14.1 Pure Strategy Nash Equilibrium

We can now define the notion of pure strategy of Nash equilibrium. Given an extensive form game $\langle G = (N, A_i, H, P, U_i)\rangle$ and a player action profile $p' = (p'_1, p'_2, \ldots, p'_n)$ is called a pure strategy Nash equilibrium if $u_i(O(p'_i, p'_{-i})) \geq u_i(O(p''_i, p'_{-i})) \quad \forall p_i \in S_i; \ \forall\, i \in N$.

If we consider the prisoner's dilemma problem, by following the above definition, it is clear that both (NC, NC) and (C, C) are pure strategy Nash equilibria.

## 6.14.2 Dominated and Dominating Mixed Strategies

If the utility obtained by adopting a strategy profile $p'$ is greater than or equal to the utility obtained by adopting another strategy profile $p''$, then the game is said to have a pure strategy Nash equilibrium. This can be extended to mixed strategies also in the context of a strategic form game $\langle N, S_i, u_i\rangle$.

**Strict dominance**

Given two mixed strategies $\phi_i, \phi_i^{'} \in \delta(S_i)$, we say that $\phi_i$ strictly dominates $\phi_i^{'}$ iff

$$u_i(\phi_i, \phi_{-i}) > u_i\left(\phi_i^{'}, \phi_{-i}\right) \forall \phi_{-i} \in \delta(S_{-i})$$

**Weak dominance**

Given two mixed strategies $\phi_i, \phi_i^{'} \in \delta(S_i)$, we say that $\phi_i$ weakly dominates $\phi_i^{'}$ iff

$$u_i(\phi_i, \phi_{-i}) \geq u_i\left(\phi_i^{'}, \phi_{-i}\right) \forall \phi_{-i} \in \delta(S_{-i})$$

and

$$u_i(\phi_i, \phi_{-i}) > u_i\left(\phi_i^{'}, \phi_{-i}\right) \text{for some } \phi_{-i} \in \delta(S_{-i})$$

**Very weak dominance**

Given two mixed strategies $\phi_i, \phi_i^{'} \in \delta(S_i)$, we say that $\phi_i$ strictly dominates $\phi_i^{'}$ iff

$$u_i(\phi_i, \phi_{-i}) \geq u_i\left(\phi_i^{'}, \phi_{-i}\right) \ \forall \ \phi_{-i} \in \delta(S_{-i})$$

Note that it can be shown that any dominant mixed strategy equilibrium is also a mixed strategy Nash equilibrium. If there exists a strictly dominant mixed strategy Nash equilibrium for any player in a game, it is always unique.

**Illustration 6.17**

If we consider the prisoner's dilemma problem with the following payoff matrix:

| | **Player 2** | |
|---|---|---|
| **Player 1** | **NC** | **C** |
| **NC** | -10, -10 | -2, -20 |
| **C** | -20, -2 | -2,-2 |

(a1)

| | **Player 2** |
|---|---|
| **Player 1** | **C** |
| **NC** | -2, -20 |
| **C** | -2,-2 |

(a2)

| | **Player 2** |
|---|---|
| **Player 1** | **C** |
| **C** | -2,-2 |

(a3)

Figure 6.3: Elimination of strictly dominated strategies from the prisoner's dilemma problem.

Since the action *NC* is strictly dominated by action *C* for player 2, he would never choose *NC*. So action *NC* can be eliminated, thus giving a reduced payoff matrix as in Figure (a2). Now, the action *NC* of player 1 which is dominated by strategy *C* can also be eliminated, leading to a single entry payoff matrix with profile *(C, C)*, which happens to be strongly dominant strategy equilibrium.

## 6.15 Review Questions

1. Write a short note on the characteristics of game theory. List the assumptions.
2. Define
   (a) Minimax principle
   (b) Maximin principle
   (c) Strategy
   (d) Saddle point
   (e) Value of the game
   (f) Dominance property
   (g) Two-person zero-sum game.
3. Discuss the importance of game theory to business decisions.
4. Write a short note on the limitation of game theory.
5. Consider the game *G* with the following payoff matrix:

| | Player *B* | |
|---|---|---|
| Player *A* | 2 | 6 |
| | 2 | $\mu$ |

6. Solve the following game, given the payoff matrix:

| | Player *Y* | | |
|---|---|---|---|
| Player *X* | 4 | 8 | 4 |
| | 0 | -6 | -4 |
| | 4 | 12 | -8 |

7. Solve the following games:

(a)

| | Player *B* | | |
|---|---|---|---|
| Player *A* | 15 | 2 | 3 |
| | 6 | 5 | 7 |
| | -7 | 4 | 0 |

(b)

| | Player *B* | | | |
|---|---|---|---|---|
| Player *A* | -25 | 25 | 0 | 35 |
| | 10 | 30 | 5 | 40 |
| | -20 | 0 | 5 | -15 |

8. Two companies $A$ and $B$ are competing for the same product. Their different strategies are given in the following payoff matrix. Find the best strategies for the players.

Player $B$

| Player $A$ | | $B_1$ | $B_2$ | $B_3$ |
|---|---|---|---|---|
| | $A_1$ | 4 | -4 | 6 |
| | $A_2$ | -6 | 10 | 2 |

9. Two players select one of the boxes "Rambo," "Mogambo," "Sambo," and "Dambo" independently. Listed are some facts about the boxers.

   (a) Rambo can beat Mogambo for a score of 8.
   (b) Mogambo can beat Dambo for a score of 6.
   (c) Dambo can beat Sambo for a score of 4.
   (d) Sambo can beat Rambo for a score of 2.

   Formulate the payoff matrix and find the optimal strategies for both the players and the value of the game.
   **Hint**: Payoff matrix.

Player $B$

| Player $A$ | | Rambo | Sambo | Dambo | Mogambo |
|---|---|---|---|---|---|
| | Rambo | 0 | -2 | 0 | 8 |
| | Sambo | 2 | 0 | -4 | 0 |
| | Dambo | 0 | 4 | 0 | -6 |
| | Mogambo | -8 | 0 | 6 | 0 |

10. Solve the following game and determine the value of the game:

Player $B$

| Player $A$ | | Strategy 1 | Strategy 2 |
|---|---|---|---|
| | Strategy 1 | 4 | 1 |
| | Strategy 2 | 2 | 3 |

11. Explain the principle and rule of dominance to reduce the size of the payoff matrix.
12. Solve the following games by dominance principle:

(a)

Player $B$

| Player $A$ | | I | II | III | IV |
|---|---|---|---|---|---|
| | 1 | -5 | 3 | 1 | 20 |
| | 2 | 5 | 5 | 4 | 6 |
| | 3 | -4 | 2 | 0 | -5 |

(b)

| | | Player B 1 | 2 | 3 |
|---|---|---|---|---|
| Player A | 1 | 8 | 5 | 8 |
| | 2 | 8 | 6 | 5 |
| | 3 | 7 | 4 | 5 |
| | 4 | 6 | 5 | 6 |

13. *A* and *B* play a game in which each has three coins: 50p, Rs. 1, and Rs. 2. Each selects a coin without the knowledge of the other's choice. If the sum of the coins is an odd amount, *A* wins *B*'s coin. If the sum is even, *B* wins *A*'s coin. Find the best strategy for each player and the value of the game.
14. Two firms are competing for business under the conditions that gain to one firm ensures loss to the other firm. Firm *A*'s payoff matrix is given below. Suggest the optimum strategies for the two firms and hence the net outcome.

| | | Player B: No advertising | Medium advertising | Heavy advertising |
|---|---|---|---|---|
| Firm A | No advertising | 10 | 5 | -2 |
| | Medium advertising | 13 | 12 | 15 |
| | Heavy advertising | 16 | 14 | 10 |

15. Solve the following games by graphical method:

(a)

| | Player B | |
|---|---|---|
| Player A | -6 | 7 |
| | 4 | -5 |
| | -1 | -2 |
| | -2 | 5 |
| | 7 | 6 |

(b)

| | Player B | | | | |
|---|---|---|---|---|---|
| Player A | 4 | -8 | 12 | -6 | 10 |
| | -6 | 8 | -8 | 2 | 0 |

16. The following matrix represents the payoff to *A* in a rectangular game between *A* and *B*. By the notion of dominance, it was shown that the game is equivalent to one represented by a $2 \times 4$ matrix which is a sub-matrix of the given matrix. Then, obtain the solution graphically.

| | Player B | | | |
|---|---|---|---|---|
| Player A | 19 | 15 | -5 | -2 |
| | 19 | 15 | 17 | 16 |
| | 0 | 20 | 15 | 5 |

# Chapter 7
# Queuing Systems

The key takeaways for the reader from this chapter are:

- Gives elements of a queuing model
- Different types of queuing models
- Different applications
- Solution to practical problems.

## 7.1 Introduction

We come across people "queuing" for many activities in daily life. It can be the issue of "ration" in a ration shop, issue of cinema tickets, issue of rail/airline tickets, etc. The arriving people are called "customers," and the person issuing the ticket is referred to as "server." There can be more than one queue and more than one server in many cases—typically rail tickets, bus tickets. If the server is free at the time of arrival of the customer, he can be serviced immediately. If there are a number of people, a waiting line and consequently waiting time come into operation. There can also be server idle time. Table 7.1 gives some of the queuing systems applications.

## 7.2 Elements of a Queuing Model

Whenever a one-day international cricket match is scheduled, one witnesses large queues at the stadium gates prior to the start of the match, to purchase tickets. The actors in this situation are the customers arriving to buy the ticket, and the service counter at the stadium or the server providing the service by selling tickets. The arrival process is represented by *the inter-arrival times*, and the service process is represented by the *service time per customer*. The inter-arrival time is the time between successive customer arrivals. The service time per customer is


DOI: 10.1007/978-81-322-1901-9_7, 

the time taken to provide the customers the expected service as desired. In the above cricket match example, the inter-arrival time may be very less during an ODI and may be slightly high during a test match. Similarly, the service rate will be high during ODI matches and will be slightly low during the test matches. In other words, the inter-arrival times and service rates are probabilistic. *Queue discipline* represents the manner in which the customers waiting in a queue are served. It may be first-come-first-serve (FCFS), service in random order (SIRO), or last-come-first-serve (LCFS). If there are more queues, then the customers join the queue where the length is small. This is known as *jockeying*. Sometimes, the customers tend to move away from the queue place on seeing its length. This is known as *balking*. If the customers wait for a long time in the queue, but have not been serviced, they may move away. This is known as *reneging*.

There are varieties of queuing models that arise from the elements of a queue. This chapter discusses these models.

**Table 7.1**

| Situation | Customer | Servant (s) |
|---|---|---|
| Ration shop | Valid ration card holder | Clerk at theshop |
| Traffic junction | Vehicles | Signals |
| Runways | Aircrafts | Runways |
| Toll gate | Vehicles | Collectors |
| Railway ticket counter | Customers | Clerk at the counter |

## 7.3 Model I: Pure Birth Model

The pure birth model considers only arrivals, and the inter-arrival time in the pure birth model is explained by the exponential distribution. Birth of babies is a classical example for the pure birth model.

Let $P_0(t)$ be the probability of no customer arrivals during a period of time $t$.

Let $P_n(t)$ be the probability of $n$ arrivals during a period of time $t$.

We have $P_0(t) = e^{-\lambda t}$,

$$P_n(t) = \frac{(\lambda t)^n e^{-\lambda t}}{n!}, n = 0, 1, 2\ldots$$

This is a Poisson distribution with mean $E\{n/t\} = \lambda t$ arrivals during a period of time $t$.

**Illustration 7.1**

An art collector travels to art auctions once a month on average. Each trip is guaranteed to produce one purchase. The time between trips is exponentially distributed. Determine the following:

(a) The probability that the collector will not purchase any art pieces in a 3-month period.
(b) The probability that the collector will not purchase more than eight art pieces per year.
(c) The probability that the time between successive trips will exceed 1 month is

**Solution**
From the data given we have,

$$\lambda = 1 \text{ trip/month}$$

(a) The probability that the collector will not purchase any art pieces in a 3-month period is calculated using the formula

$$P_n(t) = \frac{(\lambda t)^n e^{-\lambda t}}{n!}$$

$$P_n(t=3) = \frac{(3\times 1)^0 e^{-3\times 1}}{0!} = e^{-3} = 0.0497$$

(b) Similarly, the probability that the collector will not purchase more than eight art pieces per year will be

$$\begin{aligned}
P_{n\leq 8}(\text{t}=12) &= P_0(12)+P_1(12)+P_2(12)+P_3(12)+P_4(12)+P_5(12)\\
&\quad +P_6(12)+P_7(12)+P_8(12)\\
&= \frac{(12)^0e^{-12}}{0!}+\frac{(12)^1e^{-12}}{1!}+\frac{(12)^2e^{-12}}{2!}+\frac{(12)^3e^{-12}}{3!}+\frac{(12)^4e^{-12}}{4!}\\
&\quad +\frac{(12)^5e^{-12}}{5!}+\frac{(12)^6e^{-12}}{6!}+\frac{(12)^7e^{-12}}{7!}+\frac{(12)^8e^{-12}}{8!}\\
&= 0.0001+0.0007+0.00044+0.00177+0.00531\\
&\quad +0.01274+0.02548+0.04368+0.0552\\
&= 0.15503
\end{aligned}$$

Therefore, the probability will be 0.15503.

(c) The probability that the time between successive trips will exceed 1 month

$$P_0(1) = \frac{1^0\cdot e^{-1}}{0!} = e^{-1} = 0.3674$$

Therefore, the probability will be 0.3674.

**Illustration 7.2**
In the central bus stand, the arrival rate of customers is 2/min. Determine the following:

(a) The average number of arrivals during 5 min.
(b) The probability that no arrivals will occur in the next half a minute.
(c) The probability that at least one arrival will occur during the next half a minute.
(d) The probability that the time between successive arrivals is at least 3 min.

**Solution**
The problem that $\lambda = 2$ customers/min.

(a) The average number of arrivals during 5 min is calculated by

$$\lambda t = 2 \times 5 = 10 \text{ arrivals}$$

(b) Probability of no arrivals in the next 0.5 min is

$$P_0(\mathrm{t} = 0.5) = e^{-2\times 0.5} = e^{-1} = 0.3679 \text{ or } 36.79\,\%$$

(c) Probability of at least one arrival in the next 0.5 min

$$= 1 - P_0(\mathrm{t} = 0.5) = 1{-}0.3679 = 0.6321$$

(d) Probability that time between two successive arrivals is at least 3 min is

$$P_0(t = 3) = \frac{(2 \times 3)^0 \cdot e^{-2\times 3}}{0!} = e^{-6} = 0.00247$$

**Illustration 7.3**
The time between arrivals at R&R restaurant is exponential with mean 5 min. The restaurant opens for business at 11.00 a.m. Determine the following:

(a) The probability of having 10 arrivals in the restaurant at 11:12 a.m. given that there were 8 at 11:05 a.m.
(b) The probability that a new customer will arrive between 11:28 and 11:33 a.m. given that the last customer arrived at 11:25 a.m.

**Solution**
We have $\lambda = 1/5 = 0.2$ arrival/minute

(a) Between 11:05 and 11:12 a.m., we have 7 min.

The probability of 10 arrivals, given there were 8 is

$$P_2(t = 7) = \frac{(0.2 \times 7)^2 \cdot e^{-0.2\times 7}}{2!} = 0.2416$$

(b) Similarly, probability of arrival of one new customer between 11:28 and 11:33 a.m., i.e., in a time gap of 5 min is

$$P_1(t=5)=\frac{(0.2\times 5)^1\cdot e^{-0.2\times 5}}{1!}=0.3679$$

**Illustration 7.4**

Prove that the mean and variance of the Poisson distribution during an interval $t$ equals $\lambda t$, where $\lambda$ is the arrival rate.

**Solution**

We have the mean of Poisson distribution as

$$\begin{aligned}E[n/t]&=\sum_{n=1}^{\infty}\frac{n(\lambda t)^n e^{-\lambda t}}{n!}\\&=\lambda te^{-\lambda t}\sum_{n=1}^{\infty}\frac{(\lambda t)^{n-1}}{(n-1)!}\\&=\lambda te^{-\lambda t}\cdot e^{\lambda t}=\lambda t.\end{aligned}$$

Therefore, the mean of Poisson distribution equals $\lambda t$.

The variance of the Poisson distribution is given as

$$\begin{aligned}E[n^2/t]&=\sum_{n=0}^{\infty}\frac{n^2(\lambda t)^n e^{-\lambda t}}{n!}\\&=\sum_{n=1}^{\infty}\frac{n^2(\lambda t)^n e^{-\lambda t}}{n!}\\&=\lambda te^{-\lambda t}\sum_{n=1}^{\infty}\frac{n(\lambda t)^{n-1}}{(n-1)!}\\&=\lambda te^{-\lambda t}.\frac{\delta}{\delta(\lambda t)}\left\{\lambda t\sum_{n=1}^{\infty}\frac{(\lambda t)^{n-1}}{(n-1)!}\right\}\\&=\lambda te^{-\lambda t}.\frac{\delta}{\delta(\lambda t)}\left\{\lambda te^{\lambda t}\right\}\\&=\lambda te^{-\lambda t}\left\{\lambda te^{\lambda t}+e^{\lambda t}\right\}\\&=(\lambda t)^2+\lambda t\end{aligned}$$

So, the var$\{n/t\}=(\lambda t)^2+\lambda t-(\lambda t)^2=\lambda t$

Therefore, the variance of the Poisson distribution also equals $\lambda t$.

## 7.4 Model II: Pure Death Model

The pure death model contradicts the pure birth model, in the sense that only departures are considered in this model. Here, the system starts with $N$ customers at time 0, and no further arrivals are permitted to the system. The departure rate is $\mu$ customers per unit time. $P_n(t)$ is the probability of $n$ customers in the system after $t$ time units.

So, we have

$$P_n(t) = \frac{(\mu t)^{N-n} e^{-\mu t}}{(N-n)!}, n = 1, 2, \ldots N; \text{and}$$

$$P_0(t) = 1 - \sum_{\text{n}=1}^{N} P_0(t)$$

**Illustration 7.5**

Demand for an item occurs according to a Poisson distribution with mean 3 per day. The maximum stock level is 25 items, which occurs on each Monday immediately after a new order is received. The order size depends on the number of units left at the end of week on Saturday (business is closed on Sundays). Determine the following:

(a) The average weekly size of order
(b) The probability of incurring shortage when the business opens on Friday morning
(c) The probability that the weekly order size exceeds 14 units.

**Solution**

Given, $N = 25$, $\mu = 3$/day, $t = 6$ days, $\mu t = 3 \times 6 = 18$

(a) The average stock remaining after 6 days

$$E\{n/t = 6\} = \sum_{\text{n}=0}^{25} np_n(6)$$

We have

$$P_n(t) = \frac{(\mu t)^{N-n} e^{-\mu t}}{(N-n)!}$$

$$P_0(6) = \frac{(18)^{25} e^{-18}}{(25)!} = 0.0236$$

$$P_1(6) = \frac{(18)^{24} e^{-18}}{(24)!} = 0.0328$$

Similarly, we calculate all the values up to $P_{25}(6)$ and $E\{n/t = 6\} = 7.10$
The average order size $= 25 - 7.10 = 17.9 \sim 18$ items.

(b) Up to Friday from Monday, we have 4 days without considering Friday, i.e., $t = 4$.

$$\mu t = 3 \times 4 = 12$$

$$\text{So}, P_0(4) = 1 - P_0 \sum_{n=1}^{25} P_n(4) = 0.00069$$

The probability of incurring shortage when the business opens on Friday morning is 0.00069

(c) In a week, we have 6 days, i.e., $t = 6$, $\mu t = 3 \times 6 = 18$

$$P_{n>14}(6) = 1 - P_{n \le 14}(6)$$

We have

$$P_{n \le 14}(6) = P_0(6) + P_1(6) + \cdots + P_{14}(6) = 0.652$$

Therefore,

$$1 - P_{n \le 14}(6) = 1 - 0.652 = 0.348$$

The probability that weekly order size exceeds 14 units is 0.348

**Illustration 7.6**

Prove that the distribution of time between departures corresponding to the Poisson distribution in the pure death is an exponential distribution with mean $\frac{1}{\mu}$ time units.

**Solution**

We have the Poisson distribution formula for the pure death model as

$$P_n(t) = \frac{(\mu t)^{N-n} e^{-\mu t}}{(N-n)!}, n = 1, 2, \ldots N$$

We have

$$\begin{aligned} &P(\text{Time between departures} > T) \\ &= P(\text{No departures during } T) \\ &= P(N \text{ left after time } T) \\ &= P_N(T) \end{aligned}$$

Therefore, this can be represented as

$$P(t > T) = P_N(T) = \frac{(\mu t)^0 e^{-\mu t}}{(0)!} = e^{-\mu t}$$

**Illustration 7.7**

Inventory is withdrawn from a stock of 80 items according to a Poisson distribution at the rate of 5 items per day. Determine the following:

(a) The probability of 10 items being withdrawn during the first 2 days
(b) The probability that no items are left at the end of 4 days
(c) The average number of items withdrawn over a 4-day period

**Solution**

Given, $N = 80$ items, $\mu = 5$ items/day

(a) We have $t = 2$, $\mu t = 5 \times 2 = 10$ items

$$\text{So } P_{10}(2) = \frac{10^{80-10}e^{-10}}{(80-10)!} = 0.125$$

So, the probability of 10 items being withdrawn during the first 2 days $= 0.125$

(b) $t = 4$, so we have $\mu t = 5 \times 4 = 20$ items

$$\text{So } P_0(4) = 1 - \sum_{n=1}^{N} P_n(4)$$
$$= 1 - 0.99999 = 0.00001$$

The probability that no items are left at the end of 4 days is 0.00001, i.e., negligible.

(c) Again, we have $t = 4$, $\mu t = 5 \times 4 = 20$ items.

The average number of items withdrawn over a 4-day period

$$E\{n/4\,\text{days}\} = \sum_{n=0}^{80} P_n(4) \sim 60\,\text{items}$$

So, the average number of withdrawals $= 80 - 60 = 20$ items.

**Illustration 7.8**

Kumarans High School band is performing a benefit jazz concert in its 400-seat auditorium. Locals buy tickets in blocks of 10 and donate them to youth organizations. Tickets go on sale to businesses for 4 h only the day before the concert. The process of placing order for tickets follows a Poisson distribution with a mean of 10 calls/h. Any tickets remaining after the box office is closed, are sold at a discount as "rush hour tickets," an hour before the start of the concert. Determine the following:

(a) The probability that it will be possible to buy rush hour tickets
(b) The average number of rush hour tickets.

**Solution**

Given $N = 40$, $\mu = 10$ calls/h

(a) The probability that it will be possible to buy rush hour tickets will be ($t = 4$),

$$P_{n>0}(t = 4) = 1 - P_0(4)$$
$$P_0(4) = 0.5213 \quad 1 - P_0(4) = 1 - 0.5213 = 0.4787$$

(b) Average number of rush hour tickets

$$E\{n/t = 4\} = \sum_{n=0}^{40} nP_n(4) \sim 2.5 \text{ blocks}$$

Number of rush hour tickets = 2.5 × 10 = 25 tickets.

## 7.5 Model III: Generalized Poisson Queuing Model

This model considers both inter-arrival time and service time, and both these times follow exponential distribution. During the early operation of the system, it will be in the transient state. On the contrary, if the system is in operation for a long time, it attains steady state. In this model, both the inter-arrival and the service time exhibit state dependency.

Let us take

$n$ as the number of customers in the system (system refers to those customers who are waiting for service and who are being serviced)

$\lambda_n$ is the arrival rate where there are n customers in the system already

$\mu_n$ is the service rate where there are n customers in the system already

$P_n$ is the steady-state probability of n customers in the system.

All the above steady-state probabilities help in determining the different parameters for the model such as average queue length, waiting time in the system, and various measures of system's performance.

Here,

$$P_n = \left(\frac{\lambda_{n-1}\lambda_{n-2}\cdots\lambda_0}{\mu_n\mu_{n-1}\cdots\mu_1}\right)P_0, \quad n = 1, 2, \ldots \tag{7.1}$$

We can determine $P_0$ from the equation $\sum_{n=0}^{\infty} P_0 = 1$

For $n = 0$,

$$P_1 = \left(\frac{\lambda_0}{\mu_1}\right)P_0 \tag{7.2}$$

For $n = 1$,

$$\lambda_0 P_0 + \mu_2 P_2 = (\lambda_1 + \mu_1) P_1$$

Substituting the values of $P_1$, we get

$$P_2 = \left(\frac{\lambda_1 \lambda_0}{\mu_2 \mu_1}\right) P_0$$

Further simplification yields Eq. (7.1), i.e., value of steady-state probability $P_n$.

**Illustration 7.9**

Family Mart initially operated with three check counters. The scheduling of counters is given in Table 7.2. The arrival rate of customers is 12/h, and the check time is 15/h. What is the steady-state probability in checking slots?

**Table 7.2** The Scheduling of counters

| Customers | Counters Operating |
|---|---|
| 1 – 3 | 1 |
| 4 – 6 | 2 |
| 6 and above | 3 |

**Solution**

Given $\lambda = 12 = \lambda_n$

$$\mu_n = \begin{cases} \frac{60}{15} = 4/\text{h} & n = 1, 2, 3 \\ 2 \times 4 = 8/\text{h} & n = 4, 5, 6 \\ 3 \times 4 = 12/\text{h} & n = 7, 8, 9, 10.... \end{cases}$$

$$\begin{aligned} P_1 &= \tfrac{12}{4} P_0 = 3P_0 \\ P_2 &= \tfrac{12}{4} \times \tfrac{12}{4} P_0 = 9P_0 \\ P_3 &= \tfrac{12}{4} \times \tfrac{12}{4} \times \tfrac{12}{4} P_0 = 27P_0 \\ P_4 &= \tfrac{12}{4} \times \tfrac{12}{4} \times \tfrac{12}{4} \times \tfrac{12}{12} P_0 = 27P_0 \\ P_5 &= \tfrac{12}{4} \times \tfrac{12}{4} \times \tfrac{12}{4} \times \tfrac{12}{12} \times \tfrac{12}{12} P_0 = 27P_0 \\ P_6 &= \tfrac{12}{4} \times \tfrac{12}{4} \times \tfrac{12}{4} \times \tfrac{12}{12} \times \tfrac{12}{12} \times \tfrac{12}{12} P_0 = 27P_0 \\ P_n &= \tfrac{12}{4} \times \tfrac{12}{4} \times \tfrac{12}{4} \times \tfrac{12}{12} \times \tfrac{12}{12} \times \tfrac{12}{12} \times \left(\tfrac{12}{15}\right)^{n-6} P_0 = 27\left(\tfrac{4}{5}\right)^{n-6} P_0 \end{aligned}$$

Now, we have to determine $P_0$ by

$$P_0 + P_0\left\{3 + 9 + 27 + 27 + 27 + 27 + 27\left(\tfrac{4}{5}\right) + 27\left(\tfrac{4}{5}\right)^2 + 27\left(\tfrac{4}{5}\right)^3 + \cdots\right\} = 1$$

$$P_0\left\{94 + 27\left(1 + \left(\tfrac{4}{5}\right) + \left(\tfrac{4}{5}\right)^2 + \left(\tfrac{4}{5}\right)^3 + \cdots\right)\right\} = 1$$

From the formula

$$\sum_{n=1}^{\infty} X^K = \frac{1}{1-K} \quad |K| < 1,$$

we get

$$P_0\left\{94 + 27\left(\frac{1}{1-\frac{4}{5}}\right)\right\} = 1$$

$$P_0 = 0.00436 \text{ or } \frac{1}{229}$$

If 1 and 3 customers are present

$$P_1 + P_2 + P_3 = (3 + 9 + 27)\left(\frac{1}{229}\right) = 0.1703$$

**Illustration 7.10**

For the above example in Illustration 7.9, determine the probability distribution of number of open counters

**Solution**

Probability that one counter will be open can be defined as

$$P_0 + P_1 + P_2 + P_3 = (1 + 3 + 9 + 27)\left(\frac{1}{229}\right) = 0.174$$

Probability that 2 counters will be open can be defined as

$$P_4 + P_5 + P_6 = (27 + 27 + 27)\left(\frac{1}{229}\right) = 0.353$$

Probability that 3 counters will be open is

$$= P_7 + P_8 + \cdots + P_n$$
$$= 1 - \{P(\text{one counter will be open}) + P(\text{two counters will be open})\}$$
$$= 1 - (0.174 + 0.353) = 0.473$$

**Illustration 7.11**

In illustration 7.9, suppose all 3 counters are always open and that the operation is set up such that the customer will go to the first empty counter.

Determine the following:

(a) The probability that all 3 counters will be in use
(b) The probability that an arriving customer will not wait.

**Solution**

First, we can define $\mu_n$ as

$$\mu_n = \begin{cases} 4n, & n = 1,2 \\ 12, & n = 3,4,\ldots \end{cases}$$

$$P_1 = \left(\frac{12}{4}\right)P_0 = 3P_0$$

$$P_2 = \left(\frac{12}{4}\right)\left(\frac{12}{12}\right)P_0 = 3P_0$$

$$P_{n\geq 3} = \left(\frac{12}{4}\right)\left(\frac{12}{12}\right)\left(\frac{12}{15}\right)^{n-2}P_0 = 3\left(\frac{4}{5}\right)^{n-2}P_0$$

So,

$$P_0 + 3P_0 + 3P_0 + \left[3\left(\frac{4}{5}\right) + 3\left(\frac{4}{5}\right)^2 \cdots\right]P_0 = 1$$

We get $P_0 = 0.047$

(a) Probability that 3 counters are in use

$$\begin{aligned} P_{n\geq 3} &= 1 - (P_0 + P_1 + P_2) \\ &= 1 - [0.047 + (3 \times 0.047) + (3 \times 0.047)] \\ &= 0.671 \end{aligned}$$

(b) Probability that an arriving customer will not wait

$$P_{n\leq 2} = (P_0 + P_1 + P_2) = 0.329$$

## 7.6 Empirical Queuing Models

The basic queuing models can be classified into six categories using Kendall notation which uses six parameters to define a model (*P*/*Q*/*R*): (*X*/*Y*/*Z*).

The parameters of the notation are

*P* is the distribution of the arrival rate
*Q* is the distribution of the service rate
*R* refers to the number of service channels providing the service
*X* is the service discipline; it may be general, FCFS, SIRO, LCFS
*Y* is the maximum number of customers allowed to stay in the system at any point in time
*Z* is the calling source size.

## 7.7 Single Server Models

### 7.7.1 Model IV (M/M/1): (GD/∞/∞)

The features of this model are:

1. There is a single service channel providing the service
2. Arrival rate or input follows Poisson distribution
3. Service rate is exponentially distributed
4. There is no limit on the system's capacity
5. Customers are served on first-come-first-served basis.

**Symbols and Notations**

1. $\lambda$—Arrival rate of customers (numbers/hour)
2. $\mu$—Service rate (numbers/hour)
3. $T$—Mean time between arrivals = $1/\lambda$
4. $t$—Average time of servicing = $1/\mu$
5. $\rho$(rho)—Utilization factor or traffic intensity = $\lambda/\mu$
6. $\rho_0$—Idling factor of the facility, i.e., probability of providing the service right away to the customers without him having to wait = $1 - \rho$
7. $P_n$—Probability that there are $n$ customers waiting in the system for service.
8. $N$—Number of customers in the system
9. $L_q$—Length of the queue or average number of customers in the queue waiting for service
10. $L_s$—Average number of customers in the system (both at service counters and queue)
11. $W_q$—Average waiting time in the queue
12. $W_s$—Average waiting time in the system
13. $W/W > 0$—Expected waiting time of a customer who has to wait
14. $L/L > 0$—Expected length of nonempty queue

The formula list for the model $(M/M/1)$: (GD/∞/∞) is given below:

1. $P_0 = 1-(\lambda/\mu) = 1-\rho$
2. $P_n = \left(\frac{\lambda}{\mu}\right)\left(1-\frac{\lambda}{\mu}\right) = (\rho)^n \times (1-\rho)$
3. Probability of queue size greater than $n$ $(Q \geq n) = (\rho)^n$
4. $L_s = \left(\frac{\lambda}{\mu-\lambda}\right)$
5. $L_q = \left(\frac{\lambda}{\mu-\lambda}\right) \times \left(\frac{\lambda}{\mu}\right) = \left(\frac{\lambda^2}{\mu(\mu-\lambda)}\right)$
6. $W_s = \left(\frac{\lambda}{\mu-\lambda}\right) \times \left(\frac{1}{\lambda}\right) \quad L_s \times \left(\frac{1}{\mu}\right) = \left(\frac{1}{\mu-\lambda}\right)$
7. $W_q = \left(\frac{\lambda}{\mu(\mu-\lambda)}\right)$

8. $L/L > 0 = L_n = \left(\frac{\lambda}{\mu-\lambda}\right)$
9. Average waiting time in the nonempty queue $= \left(\frac{1}{\mu-\lambda}\right)$
10. Probability of an arrival waiting for $t$ mins or more $= \rho e^{-(\mu-\lambda)t}$

**Illustration 7.12**

Customer arrival at a BMTC counter follows a Poisson distribution with a mean of 160/h. The service provided to the customer also follows a Poisson distribution with a mean of 75/h. Determine the following:

(a) The average length of system
(b) Average number of customers in the queue
(c) Average time a customer spends in the system
(d) Average time a customer waits for service
(e) Probability of having 0 customers in the system
(f) Probability of having 8 customers in the system.

**Solution**

Given, $\lambda = 60/\text{h}$, $\mu = 75/\text{h}$

$$\text{The utilization factor} = \rho = \frac{\lambda}{\mu} = \frac{60}{75} = 0.80$$

(a) $L_s t = \left(\frac{\lambda}{\mu-\lambda}\right) = \left(\frac{60}{75-60}\right) = 4\,\text{customers}$

(b) $L_q = \left(\frac{\lambda^2}{\mu(\mu-\lambda)}\right) = \left(\frac{60^2}{75(75-60)}\right) = 3.2\,\text{customers}$

(c) $W_s = \left(\frac{1}{\mu-\lambda}\right) = \left(\frac{1}{75-60}\right) = 0.066\,\text{h}$

(d) $W_q = \left(\frac{\lambda}{\mu(\mu-\lambda)}\right) = \left(\frac{60}{75(75-60)}\right) = 0.053\,\text{h}$

(e) $\rho_o = 1 - \rho = 1 - 0.8 = 0.2$

(f) $\rho_8 = \rho^n(1 - \rho^n) = 0.8^8(1 - 0.8^8) = 0.139$

**Illustration 7.13**

Vehicles pass through a toll gate at a rate of 90/h. The average waiting time to pass through the gate is 36 s. The arrival and service rates follow Poisson distribution. Following a complaint with respect to long durations, the authorities are willing to install one more gate to reduce the average waiting time to pass through the toll gate to 30 s, if the idle time of the toll gate is less than 10 %, and the average queue length at the gate is more than five vehicles. Is the installation of the second gate justified?

**Solution**

Given $\lambda = 90$/h

Time to pass through the gate = 36 s

$$\mu = \frac{60 \times 60}{36} = 100/\text{h}$$

Now,

$$\rho = \left(\frac{\lambda}{\mu}\right) = \left(\frac{90}{100}\right) = 0.9$$

The average length of the queue would be

$$L_q = \left(\frac{\lambda^2}{\mu(\mu - \lambda)}\right) = \left(\frac{90^2}{100(100 - 90)}\right) = 8.1 \text{ vehicles}$$

If the average time to pass through the gate is 30 s,

$$\mu_1 = \frac{60 \times 60}{30} = 120/\text{h}$$

$$\rho = \left(\frac{\lambda}{\mu_1}\right) = \left(\frac{90}{120}\right) = 0.75$$

$$L_q = \left(\frac{90^2}{120(120 - 90)}\right) = 2.25 \text{ vehicles}$$

Idle time percentage = $1 - \rho = 1 - 0.75 = 0.25$

So, this shows that the gate will be idling 25 % of the time, but the required idle time is to be less than 10 %. So, the installation of the second gate is not justified.

**Illustration 7.14**

Arrival of cars at Keerthi Service Station follows a Poisson distribution with a mean of 6 min, and customer service time is exponentially distributed with mean of 9 min. The garage space can accommodate only three vehicles for service at a time. All the other vehicles wait for their turn. Determine the following:

(a) Probability that a car enters the garage directly without having to wait
(b) Probability that a car waits for service
(c) Expected waiting time for a car to get the service.

**Solution**

Given $\lambda = 6$ cars/h, $\mu = 9$ cars/h

(a) Probability that a car does not wait for service

$$= P_0 + P_1 + P_2 \text{(as the maximum capacity is three cars)}$$

$$= \left(1 - \frac{\lambda}{\mu}\right) + \left(1 - \frac{\lambda}{\mu}\right)\frac{\lambda}{\mu} + \left(\frac{\lambda}{\mu}\right)^2\left(1 - \frac{\lambda}{\mu}\right)$$

$$= \left(1 - \frac{\lambda}{\mu}\right)\left[1 + \frac{\lambda}{\mu} + \left(\frac{\lambda}{\mu}\right)^2\right]$$

$$= \left(1 - \frac{6}{9}\right)\left[1 + \frac{6}{9} + \left(\frac{6}{9}\right)^2\right] = \frac{19}{27} = 0.703$$

(b) Probability that a car waits for service

$$= 1 - (P_0 + P_1 + P_2) = 1 - 0.703 = 0.296$$

(c)

$$W_q = \left(\frac{\lambda}{\mu(\mu - \lambda)}\right) = \left(\frac{6}{9(9 - 6)}\right) = 0.22\,\text{h}$$

**Illustration 7.15**

The XYZ company's quality control department is managed by a single clerk, who takes on average 5 min for checking the parts of each of the machines that come for inspection. The machines arrive once in every 8 min. One hour of the machine is valued at Rs. 15, and the clerks' time is valued at Rs. 4/h. What are the average hourly queuing system costs associated with the quality control department?

**Solution**

Given

$$\lambda = \frac{1}{4}/\text{min} = \frac{15}{92}/\text{h}$$

$$\mu = \frac{1}{5}/\text{min} = 12/\text{h}$$

$$W_s = \left(\frac{1}{\mu - \lambda}\right) = \left(\frac{1}{12 - \frac{15}{92}}\right) = \frac{2}{9}\,\text{h}$$

If a machine stands in a queue, the associated costs would be Rs. $\left(15 \times \frac{2}{9}\right)/\text{h} = \text{Rs.}\ \frac{10}{3}$

The arrival rate $= \left(\frac{15}{2}\right)/\text{h}$. Costs associated with this would be Rs. $\left(\frac{10}{3} \times \frac{15}{2}\right)/\text{h} = \text{Rs.}\ 25$

The costs associated with clerk (from data) = Rs. 4

The total costs = Rs. 25 + Rs. 4 = Rs. 29/h

### 7.7.2 Model V: (M/M/1): (GD/N/∞)

The features of this model are:

1. Single service channel to provide the service
2. Both arrival rate and service rate follow Poisson distribution
3. Number of customers allowed in the system cannot exceed $N$ at any point of time.

The formula list is given below:

1. (a)

$$P_N = \frac{\left(1-\frac{\lambda}{\mu}\right)}{\left(1-\frac{\lambda}{\mu}\right)} \times \left(\frac{\lambda}{\mu}\right)^N, \frac{\lambda}{\mu} \neq 1, N = 0, 1, 2, \ldots N$$

$$= \frac{(1-\rho)}{(1-\rho)^{N+1}} \times (\rho)^N, \rho \neq 1, N = 0, 1, 2, \ldots N$$

(b) If $\rho = 1$, i.e., 100 % utilization,

$$P_n = \frac{1}{N+1}$$

2. Effective arrival rate of customers ($\lambda_e$)

$$\lambda_e = \lambda(1-P_N) = \mu(L_s-L_q)$$

3. (a)

$$L_s = \frac{\left(\frac{\lambda}{\mu}\right)[1-(N+1)(\rho)^N + N(\rho)^{N+1}]}{(1-\rho)(1-\rho)^{N+1}}, \text{ if } \rho \neq 1$$

$$= \frac{(\rho)[1-(N+1)(\rho)^N + N(\rho)^{N+1}]}{(1-\rho)(1-\rho)^{N+1}}, \text{ if } \rho \neq 1$$

(b) If $\rho = 1$, $L_s$ = N/2

4. $L_q = L_s - \frac{\lambda_e}{\mu} = L_s - \frac{\lambda(1-P_N)}{\mu}$
5. $W_s = \frac{L_s}{\lambda(1-P_N)}$
6. $W_q = \frac{L_q}{\lambda_e} = \frac{L_q}{\lambda(1-P_N)} = W_s - \frac{1}{\mu}$

**Illustration 7.16**

Vehicles arrive at E. H. Dhaba with a mean arrival rate of 15/h and service rate is 18/h. The arrival and service rates follow Poisson distribution. The parking space pointing toward the road can accommodate only three vehicles. Obtain all the parameters for the above data.

**Solution**

Given $\lambda = 15/\text{h}, \quad \mu = 18/\text{h} \quad N = 3$

$$\rho = (\lambda/\mu) = 15/18 = 0.83$$

1. $P_N = \left(\frac{1-\rho}{1-\rho^{N+1}}\right) \times \rho^N = \left(\frac{1-0.83}{1-0.83^{3+1}}\right) \times 0.83^3 = 0.185\,\text{vehicles}$
2. $\lambda_e = \lambda(1-\rho_N) = 15(1-0.185) = 12.24$ vehicles per hour
3. $$L_s = \frac{(\rho)[1-(N+1)\rho^N + N\rho^{N+1}}{(1-\rho)(1-\rho^{N+1})}$$ $$= \frac{(0.83)[1-(3+1)0.83^3 + 3\times 0.83^{3+1}}{(1-0.83)(1-0.83^{3+1})} = 1.26\ \text{vehicles}$$
4. $L_q = L_s - \left(\frac{\lambda(1-P_N)}{\mu}\right) = 1.26 - \left(\frac{15(1-0.185)}{18}\right) = 0.58\,\text{vehicles}$
5. $W_s = \left(\frac{L_s}{\lambda(1-P_N)}\right) = \left(\frac{1.26}{15(1-0.185)}\right) = 0.103\,\text{h}$
6. $W_q = \left(\frac{L_q}{\lambda(1-P_N)}\right) = \left(\frac{0.58}{15(1-0.185)}\right) = 0.047\,\text{h}$

**Illustration 7.17**

The final assembly of electric generators at GK Pvt. Ltd. is produced at Poisson rate of 10 generators/h. These are then conveyed on a belt to the inspection department for final testing. The belt can hold a maximum of 7 generators. An electronic sensor will automatically stop the conveyor, once it is full, preventing the final assembly of department from assembling more units until a space becomes available. The time to inspect the generators is exponentially distributed, with a mean of 15 min.

(a) What is the probability that the final assembly department will stop production?
(b) What is the average number of generators on the conveyor belt?
(c) The production engineer claims that interruptions in the assembly department may be reduced by increasing the capacity of the belt. The engineer, in fact, claims the capacity can be increased to the point where the assembly department can operate 95 % of the time without interruption. Is this justifiable?

**Solution**

Given $\lambda = 10\ \text{generators/h} \quad \mu = 60/15 = 4\ \text{generators/h} \quad N = 7+1 = 8$

(a)
$$P_8 = \left(\frac{1-\rho}{1-\rho^{N+1}}\right) \times \rho^N = \left(\frac{1-\left(\frac{10}{4}\right)}{1-\left(\frac{10}{4}\right)^9}\right) \times (10/4)^9 = 0.6$$

(b)
$$L_q = L_s - \left(\frac{\lambda(1-P_N)}{\mu}\right) = 6.34 \text{ generators}$$

(c) Let us assume $K =$ capacity of the belt. Thus, $N = C + 1$. The assembly department remains operating until an empty space is found on the conveyor belt. The probability of this event happening is

$$\begin{aligned} P\{\text{empty space found}\} &= P_0 + P_1 + P_2 + \cdots + P_K \\ &= \frac{1-\rho}{1-\rho^{k+2}} \sum_{n=1}^{\infty} \rho^N \\ &= \frac{1-\rho}{1-\rho^{C+2}} = \frac{1-\rho^{C+1}}{1-\rho} \\ &= \frac{1-\rho^{k+1}}{1-\rho^{k+2}} \end{aligned}$$

$$\begin{aligned} \lim_{k\to\infty} \frac{1-\rho^{K+1}}{1-\rho^{K+2}} &= \lim_{k\to\infty} \frac{-(K+1)\rho^K}{-(K+2)\rho^{K+1}} \\ &= \lim_{k\to\infty} \frac{(K+1)}{(K+2)} \frac{1}{\rho} \\ &= \lim_{k\to\infty} \frac{(1+1/K)}{(1+2/K)} \frac{1}{\rho} = \frac{1}{\rho} \end{aligned}$$

We have $\rho = 10/4$, $1/\rho = 0.4$ So,

$$\lim_{k\to\infty} (P_0 + P_1 + P_2 + \cdots + P_k) = 1/\rho = 0.4.$$

This implies probability of finding an empty space on the belt cannot exceed 0.4 regardless of the length of the belt. Hence, 95 % utilization of the assembly department would not be possible, i.e., not justifiable. The result desired could be accomplished by increasing $\mu$ or decreasing $\lambda$.

**Illustration 7.18**

The probabilities $P_N$ of $N$ customers in the system for an $(M/M/1)$: $(GD/5/\infty)$ are given in Table 7.3.

**Table 7.3** The probabilities $P_N$ of $N$ customers

| $N$ | $P_N$ |
|---|---|
| 0 | 0.399 |
| 1 | 0.249 |
| 2 | 0.156 |
| 3 | 0.097 |
| 4 | 0.061 |
| 5 | 0.038 |

The arrival rate is 5 customers/h, and the service rate is 8 customers/h. Compute

(a) Probability that an arriving customer will not be able to enter the system
(b) Rate at which arriving customers will not be able to enter the system
(c) Expected number of customers in the system
(d) Average waiting time in the queue.

**Solution**
Given $\lambda = 5$ customers/h, $\mu = 8$ customers/h.

(a) A customer cannot enter the system when $N \leq 4$.

$$\begin{aligned} P_{N \leq 4} &= P_0 + P_1 + \cdots + P_4 \\ &= 0.962 \end{aligned}$$

(b) The rate at which the customers will not be able to enter the system can be defined as

$$\begin{aligned} \lambda_R &= \lambda \times P_5 \\ P_5 &= 1 - P_{N \leq 4} = 1 - 0.962 = 0.038 \\ \lambda_R &= 5 \times 0.038 = 0.19 \text{ customers/h} \end{aligned}$$

(c) Since all $P_N$ are known, $L_S$ can be defined as $\sum_{N=0}^{5} NP_N$

$$\begin{aligned} L_S &= 0 \times 0.399 + 1 \times 0.249 + 2 \times 0.156 + 3 \times 0.097 + 4 \times 0.061 + 5 \times 0.038 \\ &= 1.286 \end{aligned}$$

(d)

$$
\begin{aligned}
W_q &= \frac{L_q}{\lambda(1-P_N)} = W_s - \frac{1}{\mu}\\
W_s &= \frac{L_s}{\lambda_e}\\
\lambda_e &= (1-P_N) = 5(1-0.038) = 4.81 \text{ customers/h}\\
W_s &= \frac{1.286}{4.81} = 0.2675\,\text{h}\\
W_q &= 0.2675 - (1/8) = 0.1424\,\text{h}
\end{aligned}
$$

**Illustration 7.19**

For a (*M/M/*1): (*GD/N/*$\infty$), show that $\lambda = \mu(L_s - L_q)$

**Solution**

We have $W_s = W_q + \frac{1}{\mu}$

Multiplying $\lambda_e$ on both sides

$$\lambda_e W_s = \lambda_e W_q + \frac{\lambda e}{\mu}$$

So, we get

$$
\begin{aligned}
L_s &= L_q + \tfrac{\lambda e}{\mu} \text{ or}\\
\lambda_e &= \mu(L_s - L_q)
\end{aligned}.
$$

## 7.7.3 *Model VI* (M/M/*1*): (GD/N/N); (N > *1*)

The features of this model are:

1. Arrival and service rates follow Poisson distribution
2. One service channel provides the service
3. Number of customers in the system cannot exceed *N* at any point of time
4. Source size is also *N* in this case.

The formula list is given below.

1. Probability of $n$ customers in the system ($P_n$)

$$P_n = N_{Cn}(\rho)^n \cdot P_0 \quad 0 \le n \le c$$
$$= N_{Cn}\frac{n!\rho^n}{C!C^{n-c}} \cdot P_0 \quad C \le n \le N$$

2. $P_0 = \left\{\sum_{n=0}^{C} N_{Cn}\rho^n + \sum_{n=c+1}^{N} N_{Cn}\frac{n!\rho^n}{C!C^{n-c}}\right\}^{-1}$
3. $L_q = N - \left\{1 + \frac{1}{\rho}\right\}(1 - P_0)$
4. $\lambda_e = \mu(1 - P_0)$
5. $W_s = \left(\frac{L_s}{\lambda_e}\right)$
6. $L_s = \left(\frac{N\rho - (1 - P_0)}{\rho}\right) = N - \left(\frac{1 - P_0}{\rho}\right)$
7. $W_q = \left(\frac{L_q}{\lambda_e}\right)$

**Illustration 7.20**

Cars arrive at a service station situated in a remote location at the rate of 4/h. The service station has only one adroit mechanic who services at the rate of 3 cars/h. At present, there are five cars in the workshop. Find all the parameters of this model.

**Solution**

Given, $\lambda = 4$ cars/h, $\mu = 3$/h, $N = 5$, $C = 1$, $\rho = \frac{\lambda}{\mu} = \frac{4}{3} = 1.33$,

1.

$$P_6 = \left[\sum_{n=0}^{C} N_{Cn}\rho^n + \sum_{n=c+1}^{N} N_{Cn}\frac{n!\rho^n}{C!C^{n-c}}\right]^{-1}$$
$$= \left[\sum_{n=0}^{1} 5_{Cn}1.33^n + \sum_{n=1+1}^{45} 5_{Cn}\frac{n!1.33^n}{1!1^{n-1}}\right]^{-1}$$
$$= 0.00188$$

2. $L_s = N - \frac{1 - P_0}{\rho} = 5 - \frac{1 - 0.00188}{1.33} = 4.24$ cars
3. $L_q = N - \left(1 - \frac{1}{\rho}\right)(1 - P_0) = 5 - \left(1 - \frac{1}{1.33}\right)(1 - 0.00188) = 4.74$ cars
4. $\lambda_e = \mu(1 - P_0) = 3(1 - 0.00188) = 2.99$
5. $W_q = L_q/\lambda_e = 4.74/2.99 = 1.58$ h; $6W_s = L_s/\lambda_e = 4.24/2.99 = 1.41$ h.

## 7.8 Multiple Server Models

In this case, there will be $c$ counters or servers in parallel to serve the customer. The customer has the option of choosing the server that is free, or that has less number of people waiting for service. The mean arrival rate ($\lambda$) follows Poisson distribution, and the mean service rate ($\mu$) follows exponential distribution. In this book, we deal with two multiple server models.

### 7.8.1 *Model VII* (M/M/C): *(GD/∞/∞)*

The features of this model are:

1. Arrival rate and service rate follows Poisson distribution
2. There are $C$ serving channels
3. Infinite number of customers permitted in the system
4. *GD* stands for general discipline servicing.

The formula list is given below.

1. $\begin{cases} P_n = \frac{\rho^n}{n!} & \text{for } 0 \leq n \leq c \\ P_n = \frac{\rho^n}{C^{n-c}C!}P_0 & \text{for } n > c \end{cases}$
2. $P_0 = \left\{ \sum_{n=0}^{C-1} \frac{\rho^n}{n!} + \frac{\rho^C}{C!\left(1-\frac{\rho}{c}\right)} \right\}^{-1}$
3. $L_q = \frac{C \cdot \rho}{(C-\rho)^2} P_c$
4. $L_s = L_q + \rho$
5. $W_q = \frac{L_q}{\lambda}$
6. $W_s = W_q + \frac{1}{\mu}$

Morse (1958) shows that for (M/M/C): (GD/∞/∞)

$$L_q = \frac{\rho}{C - \rho} \text{ as } \frac{\rho}{C} \to 1$$

**Illustration 7.21**

During the late night hours, Cauvery petrol bunk in a highway experiences lorry traffic following Poisson distribution, with a mean of 12/h. All the lorries are fueled to the maximum capacity of tank. The fueling rate follows exponential distribution with a mean of 8/h. The crew size at the petrol bunk is 4. Determine all the parameters for the above model. Also find $P_0$, $P_2$, and $P_3$.

**Solution**
Given $\lambda = 12/\text{h}$, $\mu = 8/\text{h}$, $C = 4$.

$$\rho = \left(\frac{\lambda}{\mu}\right) = \left(\frac{12}{8}\right) = 1.5$$

$$P_0 = \left\{\sum_{n=0}^{C-1} \frac{\rho^n}{n!} + \frac{\rho^C}{C!\left(1 - \frac{\rho}{c}\right)}\right\}^{-1}$$

$$= \left\{\sum_{n=0}^{3} \frac{1.5^n}{n!} + \frac{1.5^4}{4!\left(1 - \frac{1.5}{4}\right)}\right\}^{-1}$$

$$= \left\{\frac{1.5^0}{0!} + \frac{1.5^1}{1!} + \frac{1.5^2}{2!} + \frac{1.5^3}{3!} + 0.3375\right\}^{-1}$$

$$= \{1 + 1.5 + 1.125 + 0.5625 + 0.3375\}^{-1} = 0.2227$$

From thc formula $P_n = \frac{\rho^n}{n!} P_0$, for $0 \leq n \leq c$, we can compute $P_2$ and $P_3$.

$$P_2 = \frac{1.5^2}{2!} \times 0.2227 = 0.2505$$

$$P_3 = \frac{1.5^3}{3!} \times 0.2227 = 0.1252$$

Now, we can find all other parameters' values

$$L_q = \frac{\rho^{C+1}}{(C-1)! \times (C-\rho)^2} P_0 = \frac{1.5^5}{3! \times (2.5)^2} \times 0.2227 = 0.045 \sim 1 \text{ lorry}$$

$$L_s = L_q + \rho = 1 + 1.5 = 2.5 \sim 2 \text{ vehicles or 3 lorries}$$

$$W_q = \frac{L_q}{\lambda} = \frac{0.045}{12} = 0.00375\,\text{h}$$

$$W_s = W_q + \frac{1}{\mu} = 0.1287\,\text{h}$$

**Illustration 7.22**
Rextronics is a new entrant into the IT market. Lack of financial resources forces the company to get along with a number of old machines, where breakdown is a good possibility. These breakdowns happen with a mean of 5/h, which follows Poisson distribution and costs the company Rs. 175/h. The company has to employ a few adroit workers to repair the machines fast, at the rate of 4/h. Commensurate with the level of adroitness, the workers charge Rs. 75/h. How many of these workers must be employed by Rextronics in order to incur minimum costs?

**Solution**

Given $\lambda = 5/\text{h}$, $\mu = 4/\text{h}$

Breakdown cost = Rs. 175/h

Labor cost = Rs. 75/h

$$\rho = \frac{5}{4} = 1.25$$

Since the arrival rate is higher than the service rate, one worker will not be sufficient.

So $C_{\min} = 2$.

(a) If

$$C = 2$$

$$P_0 = \left\{ \sum_{n=0}^{C-1} \frac{\rho^n}{n!} + \frac{\rho^C}{C!\left(1 - \frac{\rho}{c}\right)} \right\}^{-1}$$

$$= \left\{ \sum_{n=0}^{1} \frac{1.25^n}{n!} + \frac{1.25^2}{C!\left(1 - \frac{1.25}{2}\right)} \right\}^{-1}$$

$$= \{1 + 1.25 + 2.08\}^{-1} = 0.230$$

The length of the queue will be

$$L_q = \frac{\rho^{C+1}}{(C-\rho)^2 (C-1)!} = P_0 = \frac{1.25^3}{(0.75)^2 (1)!} = 0.230 = 0.7818$$

When $C = 2$

The breakdown costs will be $L_q \times 175 = 0.798 \times 175 =$ Rs. 140

Labor costs = Rs. 75 × 2 workers = Rs. 150

Total = Rs. 290

(b) If

$$C = 3$$

$$P_0 = \left\{ \sum_{n=0}^{C-1} \frac{\rho^n}{n!} + \frac{\rho^C}{C!\left(1 - \frac{\rho}{c}\right)} \right\}^{-1}$$

$$= \left\{ \sum_{n=0}^{2} \frac{1.25^n}{n!} + \frac{1.25^3}{C!\left(1 - \frac{1.25}{3}\right)} \right\}^{-1}$$

$$= \{1 + 1.25 + 0.7812 + 0.558\}^{-1} = 0.2786$$

The length of the queue will be

$$L_q = \frac{\rho^{C+1}}{(C-\rho)^2(C-1)!} = P_0 = \frac{1.25^4}{(1.75)^2(2)!} = 0.2786 = 0.0671$$

When $C = 3$
The breakdown costs will be $L_q \times 175 = 0.0671 \times 175 =$ Rs. 11.75
Labor costs = Rs. 75 × 3 workers = Rs. 225
Total = Rs. 236.75

(c) If

$$C = 4$$

$$P_0 = \left\{ \sum_{n=0}^{C-1} \frac{\rho^n}{n!} + \frac{\rho^C}{C!\left(1-\frac{\rho}{c}\right)} \right\}^{-1}$$

$$= \left\{ \sum_{n=0}^{3} \frac{1.25^n}{n!} + \frac{1.25^4}{C!\left(1-\frac{1.25}{4}\right)} \right\}^{-1}$$

$$= \{1 + 1.25 + 0.7812 + 0.3255 + 0.1479\}^{-1} = 0.2853$$

The length of the queue will be

$$L_q = \frac{\rho^{C+1}}{(C-\rho)^2(C-1)!} = P_0 = \frac{1.25^5}{(3.25)^2(3)!} = 0.2853 = 0.0446$$

When $C = 4$
The breakdown costs will be $L_q \times 175 = 0.0446 \times 175 =$ Rs. 7.81
Labor costs = Rs. 75 × 4 workers = Rs. 300
Total = Rs. 307.81 (Table 7.4).

**Table 7.4** The optimal strategy

| Workers (number) | Costs (Rs.) |
|---|---|
| 2 | 290 |
| 3 | 236.75 |
| 4 | 307.81 |

We see that employing four workers increases costs to the company. So, employing three workers would be the optimal strategy (Table 7.4).

**Illustration 7.23**
Determine the minimum number of parallel servers needed in each of the following (Poisson arrival/departure) situations to guarantee the operation of queuing situation will be stable (i.e., queue length will not grow indefinitely)

(a) Customers arrive every 5 min and service rate is 10/h
(b) The average inter-arrival time is 2 min and the average service time is 6 min
(c) The arrival rate is 30/h, and the server rate per server is 40/h.

**Solution**

(a) $\lambda = 60/5 = 12/\text{h} \quad \mu = 10/\text{h}$

$$C > \lambda/\mu = 1.2 \rightarrow C \geq 2.$$

This implies minimum number of services is 2.
(b) $\lambda = 60/2 = 30/\text{h} \quad \mu = 60/6 = 10/\text{h}$

$$C > \lambda/\mu = 3 \rightarrow C \geq 4.$$

This minimum number of servers is 4.
(c) $\lambda = 30/\text{h} \quad \mu = 40/\text{h}$

$$C > \lambda/\mu = 0.75 \rightarrow C \geq 1.$$

This minimum number of servers is 1.

## *7.8.2 Model VIII* (M/M/C): (GD/N/$\infty$)

The features of this model are:

1. The system limit is finite and is equal to $N$
2. The arrival rate and service rate follow Poisson distribution
3. The maximum queue length is $N - C$
4. There are $C$ service channels.

Here $\lambda_e < \lambda$, because of $N$
The generalized model can be defined as

$$\lambda_n = \begin{cases} (N-n)\lambda & \text{for } 0 \leq n \leq N \\ 0 & \text{for } n \geq N \end{cases}$$

$$\mu_n = \begin{cases} n\mu & \text{for } 0 \leq n \leq C \\ C\mu & \text{for } C \leq n \leq N \\ 0 & \text{for } n \geq N \end{cases}$$

The formula list is given below:

1. $P_n = \begin{cases} N_{Cn}\rho^n \cdot P_0 & \text{for } 0 \le n \le C \\ N_{Cn}\frac{n!\rho^n}{C!C^{n-C}} \cdot P_0 & \text{for } C \le n \le N \end{cases}$
2. $P_0 = \left\{ \sum_{n=0}^{C} N_{Cn}\rho^n + \sum_{n=C+1}^{N} \frac{n!\rho^n}{C!C^{n-c}} \right\}^{-1}$
3. $L_q = \sum_{n=c+1}^{N} (n - C)P_n$
4. $L_s = L_q + \frac{\lambda_e}{\mu}$
5. $\lambda_e = \mu(C - C_1), C_1 = \sum_{n=0}^{C}(C - n)P_n = \lambda(N - L_s)$
6. $W_s = \frac{L_s}{\lambda_e}$
7. $w_q = \frac{L_q}{\lambda_e}$

**Illustration 7.24**
Consider illustration 7.19. If a maximum of five lorries are allowed to stand in the petrol bunk at s time, find $P_0$, $L_q$, $L_s$, $w_q$, $w_s$.

**Solution**
$\rho = 1.5 \quad C = 4 \quad N = 5$

$$P_0 = \left\{ \sum_{n=0}^{C-1} \frac{\rho^n}{n!} + \frac{\rho^C\left(1 - \frac{\rho}{c}\right)^{N-C+1}}{C!\left(1 - \frac{\rho}{c}\right)} \right\}^{-1}$$

$$= \left\{ \sum_{n=0}^{3} \frac{1.5^n}{n!} + \frac{1.5^4\left(1 - \frac{1.5}{4}\right)^{5-4+1}}{4!\left(1 - \frac{1.5}{4}\right)} \right\}^{-1}$$

$$= \{1 + 1.5 + 1.125 + 0.5625 + 0.1318\}^{-1} = 0.2315$$

$$L_q = \frac{\rho^{C+1}}{(C-1)!(C-\rho)^2} P_0 \times \left[1 - \left(\frac{\rho}{C}\right)^{N-C} - (N-C)\left(\frac{\rho}{C}\right)^{N-C}\left(1 - \frac{\rho}{C}\right)\right]$$

$$= \frac{1.5^5}{(3)!(2.5)^2} \times 0.2315 \times \left[1 - \left(\frac{1.5}{4}\right)^1 - (1)\left(\frac{1.5}{4}\right)^1\left(1 - \frac{1.5}{4}\right)\right]$$

$$= 0.0183 \text{ lorries}$$

$$C = \sum_{n=0}^{4}(4 - n)P_n$$

$$= (4-0)P_0 + (4-1)P_1 + (4-2)P_2 + (4-3)P_3 + (4-4)P_4$$

$$= 4P_0 + 3.\frac{\rho^1}{1!}P_0 + 2.\frac{\rho^2}{2!}P_0 + 1.\frac{\rho^3}{3!}P_0 + 0$$

$$= 4 \times 0.2315 + (3 \times 1.5 \times 0.2315) + \left[2 \times \frac{1.5^2}{2} \times 0.2315\right] + \left[1 \times \frac{1.5^3}{2} \times 0.2315\right]$$

$$= 2.618 \sim 3$$

$$P_n = \frac{\rho^n}{C^{N-C}C!} P_0$$

Here, $n = N$

$$P_n = \frac{1.5^5}{4^{5-1}4!} = 0.079$$

$$L_s = L_q + \frac{\lambda_e}{\mu}$$

$$\lambda_e = \lambda(1 - P_n) = 12(1 - 0.079) = 11.05 \text{ lorries per hour}$$

$$L_s = 0.0018 + \frac{11.05}{8} = 1.399 \text{ lorries}$$

$$W_q = \frac{L_q}{\lambda_e} = \frac{0.018}{11.05} = 0.00162\,\text{h}$$

$$W_s = \frac{L_s}{\lambda_e} = \frac{1.399}{11.05} = 0.126\text{ h}$$

**Illustration 7.25**

A small engine repair shop is run by three mechanics. Here, tillers and mowers are brought for service and maintenance. The shop accepts all tillers and mowers the customers bring. The floor shop can house a maximum of 15 tillers or mowers, apart from the ones being serviced. The arrival rate is 10 min on average, and each mechanic takes half an hour to complete each job. Both arrival and service rates follow exponential distribution. Determine the following:

(a) Average number of idle mechanics
(b) Amount of business lost to competition per 10-h day because of the limited capacity of the shop.
(c) Probability that the next arriving customer will be serviced by the shop
(d) Probability that at least one of the mechanics will be idle
(e) Average number of fillers or mowers awaiting service
(f) Measure of overall productivity of the shop.

**Solution**

From the calculations, we get $L_s = 9.54 \quad L_q = 6.71$

Given $\lambda = 60/10 = 6/\text{h}$

$\mu = 60/30 = 2/\text{h}$ $N = 18 \quad C = 3$

(a) Average number of idle mechanics

$$= C - (L_s - L_q)$$
$$= 3 - (9.54 - 6.71)$$
$$= 0.17$$

(b) If the customers see the shop floor with awaiting jobs, they go elsewhere for more prompt service. So,

$$P_{18} = \frac{\rho^n}{C^{N-C}C!} P_0 = 0.0559$$

$$\lambda_{\text{lost}}(\text{jobs lost}) = 0.0559 \times 6 = 0.3354 \text{ jobs}$$

$$\text{Average jobs lost in } 10\,\text{h} = 0.3354 \times 10 = 3.354 \text{ jobs}$$

(c) $P_{n \le 17} = P_0 + P_1 + \cdots + P_{17} = 0.9441$
(d) $P_{n \le 2} = P_0 + P_1 + P_2 = 0.10559$
(e) $L_q = \sum_{n=C+1}^{N} (n - C)P_n = 6.71$ mowers
(f) Productivity measure $= \frac{L_s - L_q}{C} = \frac{9.54 - 6.71}{3} = 0.944.$

## 7.9 Review Questions and Problems

1. What are the application areas of queuing theory?
2. Discuss the terminologies used in a queuing system?
3. What is Kendall notation? How can queuing systems be classified under this?
4. The city central library in Jayanagar receives an average of 30 books per day. The arrival of these books follows Poisson distribution. Each shelf in the library can hold 85 books. Calculate the probability that more than 6 cases will be purchased in 3 months, given that one case has 4 shelves.
5. A business man deposits Rs. 5,000 in a bank every month to handle emergency situations, if any. He randomly withdraws Rs. 300 from his account, which follows a Poisson distribution with a mean value of 8 days. What is the probability that he will run out of money before the 25th day of the month.
6. Customers arrive at the box office window, being manned by a single individual, according to a Poisson input rate of 20/h. The time required to serve a customer has an exponential distribution with a mean of 90 s. What is the average waiting time of the customer?
7. Consider a self-service store with one cashier. Assume a Poisson distribution with respect to arrivals and exponential distribution with respect to service times. The store witnesses 9 arrivals every 5 min, and the cashier can serve ten customers in 5 min. Determine the following:

   (a) The probability of more than ten customers in the system
   (b) The probability of a customer has to wait in the queue for more than 2 min
   (c) The average number of customers waiting in the queue for service.

8. People arrive at a theater ticket booth according to a Poisson distribution at a rate of 25/h. Service time is constant at 2 min. Calculate the following:

   (a) The mean number in the waiting line

(b) The mean waiting time for customers
(c) The probability that an arrival will enter the cinema hall without having to wait.

9. In Bhawan cafeteria, it is observed that there is only one bearer who takes exactly 4 min to serve a cup of coffee once an order has been placed with him. If students arrive at the cafeteria at the rate of 10/h on average, how much time is one expected to spend waiting for his turn to place an order.
10. A tax consulting firm has four service stations in its offices to receive people who have problems and complaints about their income, wealth, and sales tax. Arrivals average one hundred persons in a 10-hour day. Each advisor spends an irregular amount of time servicing the arrivals which have been found to have exponential distribution. The average service time is found to be following exponential distribution. The average service time is 20 min. Calculate the following:

    (a) The average number of customers in the system
    (b) The average number of customers waiting to be served
    (c) Average time a customer spends in the system
    (d) Average waiting time for a customer
    (e) The probability that a customer has to wait before he gets service.

11. Patients arrive at a clinic according to a Poisson distribution at the rate of 30/h. The waiting lounge in the clinic is small and can accommodate not more than 14 persons at a time. The examination (service) time is exponential with a mean of 20/h. Find the following:

    (a) The effective arrival rate at the clinic
    (b) The probability that a patient will straightaway see the doctor, i.e., without having to wait in the lounge
    (c) The expected waiting time till a patient is discharged from the clinic.

12. A small post office has two open windows. Customers arrive according to a Poisson distribution at the rate of one every 3 min. However, only 80 % of them seek service at the windows. The service time per customer is exponential with a mean of 5 min. All arriving customers form one line and access available windows on an FCFS basis.

    (a) What is the probability that an arriving customer will wait in line
    (b) What is the probability that both windows are idle
    (c) What is the average length of the waiting line
    (d) Would it be possible to offer reasonable service with only one window? Explain.

13. The new Bangalore airport, when it is officially opened is said to service rural, suburban, and transit passengers effectively. Assume the arrivals follow Poisson distribution with a mean of 10, 12, and 18 passengers per hour. The time to check-in a passenger may be assumed to follow exponential

distribution with a mean of 4 min using advanced technologies. What are the number of counters you recommend to the government with respect to the number of service counters to be installed under the following conditions:

(a) The total time to check a passenger is less than 20 min
(b) The probability that the utilization of all the counters will be 90 % on all the days
(c) The probability that the utilization of the counters is 99 % on all the days.

14. Nowadays, youth want to take their cars out as soon as they get their driving licenses. They normally visit the busiest places in the city first with a lot of enthusiasm. The arrival rate of these vehicles follows Poisson distribution, with a mean of 25 cars/h. The parking lane on the road can accommodate 32 cars at a time. Once these get filled, the others have to park their cars in a side road which can accommodate 12 cars. As and when the parking lanes on the main road get freed, these cars have to be moved to the main road. The parking time (in this case, the service time) follows exponential distribution with a mean of 65 min. Calculate the following:

    (a) The probability that an arriving vehicle will have to be parked in the side road
    (b) The probability that the parking lane is free on the main road
    (c) Average number of parking lanes used on the main road
    (d) The probability of not getting a parking space both on the main road as well as the side road.

15. KSRTC employs three cashiers for better serving the customers during the holiday season. Customer arrival follows Poisson distribution with one arrival every 4 min. Customers approach the cashier immediately available. The time to serve a customer follows exponential distribution with a mean of 6 min. KSRTC is planning to construct a waiting room to accommodate the arriving customers. Your job is to find the size of waiting room, which ensures that an arrival does not wait outside with a probability of at least 0.95.
16. Ruxters bank has installed its first ATM counter. It expects a Poisson arrival with a mean of 15 customers per hour. The time needed to service the customer is expected to follow exponential distribution with a mean of 8 min. There is a small standing space in front of the ATM counter which accommodates not more than 9 people. Calculate the following:

    (a) The probability that a customer has to wait to use the ATM machine
    (b) The probability that the customer does not witness any queue in front of the ATM counter
    (c) The average number of customers in the standing space in front of the counter.

# Part II
# Engineering Management

# Chapter 8
# Introduction to Engineering Management

Learning Objectives:

- Define Engineering Management
- Understand the various roles of a manager
- Understand the origin of Engineering management
- Briefly understand the various management thoughts.

## 8.1 Introduction

In this chapter, we discuss the definition of Engineering and Management and synthesize both to get the definition of Engineering Management. We also discuss the historical development of engineering management.

## 8.2 Engineering as a Profession

Profession has been defined in Webster's third international dictionary as:

> A calling requiring specialized knowledge and often long and intensive preparation, including instruction in skills and methods as well as in the scientific, historical or scholarly principles underlying such skills and methods, maintaining by force of organization or concerted opinion high standards of achievement and conduct, and committing its members to continued study and to a kind of work which has for its prime purpose the rendering of a public service (emphasis added)

Engineer's council for Professional Development defines "Engineer" as

> The profession in which a knowledge of mathematical and natural sciences gained by study, experience, and practice is applied with judgment to develop ways to utilize, economically, the materials and forces of nature for benefit of mankind [1]


DOI: 10.1007/978-81-322-1901-9_8, 

Engineering is differentiated from other disciplines by the need to apply quantifiable principles. Academic knowledge, practical training, experience, and work-study are all required to become an Engineer.

Initially, in India, Engineering started with civil, followed by mechanical and electrical branches. Now, we have a number of branches of engineering education such as aeronautics, electrical communication engineering, telecommunications, and information technology. Experience and training help determine an Engineer's actual specialty. The next generation of engineering disciplines is characterized by flexibility and interpersonal skills.

According to estimates, every year the state of Karnataka produces more than 30,000 engineers in various disciplines. This is volume increase considering that the figure was a few hundreds in the 1950s. The figures show the need and also the employment opportunities that are being generated over the years.

## 8.3 Management

According to McFarland,

> The word manage seems to have come into English usage directly from the Italian maneggiare, meaning, 'to handle' especially to handle or train horses. It traces back to the Latin word manus, 'hand'. In early sixteenth century manage was gradually extended to the operation of war and used in general sense of taking control, taking charge or directing. Management was originally a noun used to indicate the process for managing, training, or directing. It was first applied to sports, then to housekeeping, and only later to government and business [2].

Further, McFarland identifies four important uses of the word—management, viz: (1) an organization or administrative process; (2) a science, discipline, or act; (3) the group of people running an organization, and (4) an occupational career.

Now "management" is one of the most ubiquitous and misused words. It is tagged on to practically everything, viz., waste management, figure management, etc.

## 8.4 Management Levels

In an organization, generally there are three levels of management—First-line (junior), middle, and senior (top) managers.

**First-line or Junior Managers** have typical designations of foreman, supervisor, and section head. They generally supervise non-managers and are responsible for carrying out the plans and objectives of higher management using personal and other resources given to them.

**Middle-level managers** carry titles such as production manager, division head, chief manager etc. Middle level in any organization is the key to its success or failure. They interact with the top management to get guidance and get the actual execution or plans done through the first-line managers.

**Top-level managers** have typical titles such as Managing Director, Chief Executive Officer, and Chief Operating Officer. They report to the Board of Directors and are responsible for defining the mission, character, and objective of the enterprise. The top management in an organization takes strategic policy decisions.

Figure 8.1 presents the Katz model [3], which gives the skills required by a manager. Technical skills (engineering, accounting, or word processing) are required most by the lowest level. Interpersonal skills are required at each level since it is through the efforts of other persons that goals can be achieved. Conceptual skills are required most at the top level since they should discern the critical factors that determine an organization's success or failure.

## 8.5 Managerial Roles

According to Henry Mintzberg [4], the roles played by the manager are interpersonal, informational, and decisional.

Interpersonal roles can be:

- Figure head required for ceremonial functions such as welcoming dignitaries, signing documents, and participating in important functions. These roles should not be delegated since the event itself loses significance, if done.
- Leader required for selecting, guiding, and motivating subordinates.
- Liaison required for horizontal relationships with peers and people in other organizations. These are built and nurtured for mutual assistance—referred to as networking in the present-day context.

Informational roles can be:

- Monitor involves information gathering about both internal and external events. This is done by internal review of activities, reports, attending professional meetings, etc.—sometimes referred to as function of gatekeeper.
- Disseminator involves transmission of information internally to subordinates, superiors, and peers so that everyone has the information to do his/her job. If handled well, it can strengthen the manager's formal authority.
- Spokesman (or spokesperson)—carried out by the top management. Involves giving information about organization to the external groups—press, public, etc.

Decisional roles can be:

- Entrepreneurial: to initiate change, take risk, and transform ideas into products.
- Disturbance handler: to deal with unforeseen problems or crisis and resolve them. He should know that penalty is the least effective and only one of the mechanisms for handling disturbances.

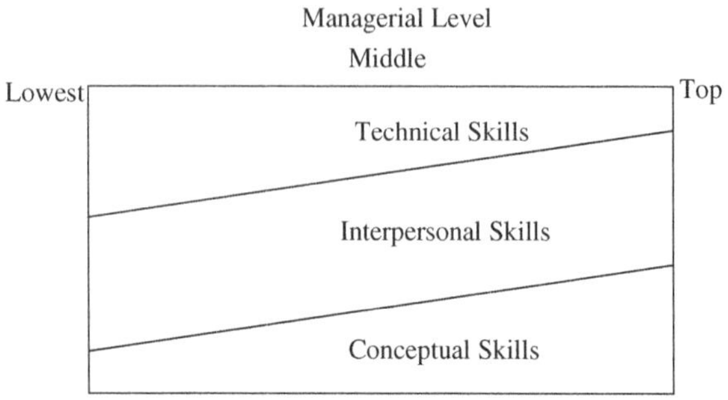


**Fig. 8.1** Skills required versus management level

- Resource allocator: to allocate resources—men, money, materials, and equipment, where they will provide the greatest benefit to the organization.
- Negotiator: to bargain with suppliers or customers, or internal persons (subordinates, superiors, or peers) to extract agreements/deals favorable to the organization.

## 8.6 Functions of Managers

The French mining engineer, Fayol [5] divided management activities into "5" elements: planning, organizing, command, coordination, and control called "Functions of Managers," the elements have proven durable over the years. In today's society, command has been replaced by "leading," "motivating," or "actuating" and subsumes coordination also. We can presently define functions of managers as follows:

- Planning involves mission selection, action to achieve them, decision making, and choosing future courses of action from among alternatives.
- Organizing involves establishing an organizational structure (includes staffing) with roles of people defined and filling them.
- Leading influencing people to strive highly toward achievement of the objectives of organization and group goals. It involves interpersonal skills.
- Controlling refers to measuring and correction of activities of subordinates to conform to plans.

## 8.7 Management: Art or Science

Management such as engineering or any other profession has a body of specialized knowledge. This forms the "science" portion. In addition, it calls for skills of interpersonal, informational, and decisional roles. This in essence is the "art" portion. Hence, Management is referred to both as art and science.

## 8.8 Engineering Management

According to Babcock,

> The engineering manager is distinguished from other managers because he (or she) possesses both an ability to apply engineering principles and a skill in organizing and directing people and projects. He is uniquely qualified for two types of jobs: the management of technical functions (such as design or production) in almost any enterprise, or the management of broader functions (such as marketing or top management) in a high-technology enterprise [6].

Today's world is characterized by high-technology businesses. These call for extensive planning so as to recognize and resolve uncertainties in the manufacture of product/s. An engineer makes himself (or herself) the obvious choice to evaluate the critical technical factors. They can also command the respect, confidence, and loyalty of subordinates. Further, his (or her) understanding of management of broader functions such as marketing or strategy makes him (or her) an invaluable asset to a high-technology enterprise.

According to Heilmeier, an electrical engineer and president of Bell core:

> Competition is global, and ability to compete successfully on this scale is fostered by corporate leaders who: Really understand the business;
>
> Understand both the technology that is driving the business today and the technology that will change the business in the future;
>
> Treat research and development as an investment to be nurtured, rather than an expense to be minimized;
>
> Spend more time on strategic thinking about the future as they rise higher in the corporation;
>
> Are dedicated to solving a customer's problem or satisfying a need, which is how I would define true marketing as opposed to sales; and
>
> Place a premium on innovation [7].

The engineering management education program is represented by Babcock as in Fig. 8.2. As can be seen, Engineering Management Education can make an engineer holistic from the business angle.

## 8.9 Origin and Growth of Engineering Management

According to records, it was in ancient Mesopotamia, lying just north and west of Babylon, that the temples developed an early concept of a "corporation" or a group of temples under a common body of management. Flourishing as early as 3000 B.C. as temple management operated activities, an administrative high-priest coordinated the secular activities of the organization. Records were kept on clay tablets, plans made, labor divided, and work supervised by a hierarchy of officials [8].

The pyramids of Egypt, the Great Wall of China, and the canal system of irrigation developed by Indian kingdoms, the latest being that of the Vijayanagar

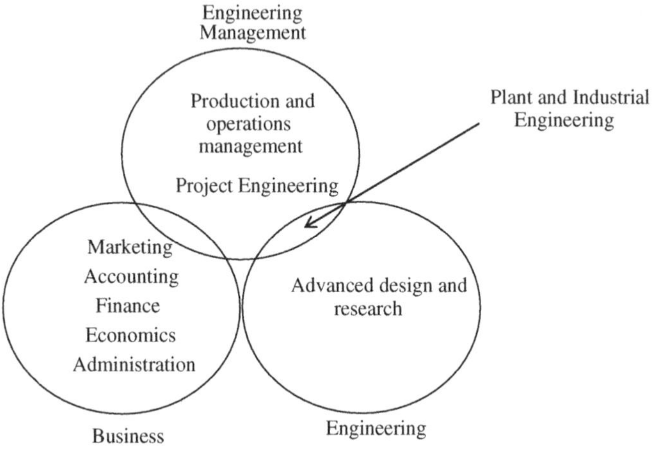


**Fig. 8.2** Engineering management Education Program *Source* Daniel L. Babcock, "BS and MS program in," Engineering Education, November 1973, p 102

Empire, makes one marvel at the engineering skills exhibited with very few tools then available.

The industrial revolution of the eighteenth century resulted in the factory revolution—textiles being very predominant. Factories started posing their problems with respect to labor, working conditions, hygiene, and sanitation. After which were the studies of scientific management by Fayol, Weber, followed by Hawthorne Experiments in Western electric (now AT&T) in the 1920s and early 1930s.

The first engineering school was probably established in France in 1747 when (Jean Rodolphe) Perronet, engineer to King Louis XV, set up his staff as a school. This group was later chartered in 1775 under the official name Ecole des pontset Chaussees (School of Bridges and Roads). Other early schools were Bergakadamie at Freiburg in Saxony (1766), Ecole polytechnic in Paris (1794), Polytechnic institute in Vienna (1815), Royal polytechnic of Berlin (1821), and University college of London (1840).

Early engineering schools to come into existence were Union College (1845), Harvard, Yale, and Michigan (1847).

## 8.10 Management Thoughts

The different management thoughts (philosophies) that have evolved over the years can be grouped under six broad streams.

- Universals of Management
- Scientific Management

**Table 8.1** Streams of management thought–universals of management and scientific management

| Universals of management | Scientific management |
|---|---|
| Fayol (1916): first complete theory of management, functions of management, principles of management, and need for management to be taught in schools | Taylor (1900): scientific management, systems applications, need for cooperation between labor and management, time study, emphasis on management's job, and emphasis on research |
| Follet (1930) followed people-oriented ideas. She said that human potential remains potential until harnessed by group association | The Gilbreths (1900): the science of motion study |
| Mooney and Reill (1931): principles of organization recognized as universal | Gantt (1901): task and bonus system, humanistic approach to labor, gantt charts, and management responsibility for training workers |
| Holden, Fish, and Smith (1948): development of charting, organizational charts—classification of company major divisions | Cooke (1919): diverse applications of scientific management |
| Urwick (1943) pointed out the existence of similarities between physical and social sciences and correlated them. Consolidated the managerial principles developed by others | |
| Davis: He placed emphasis on definitions and classifications and the explicit statement of well over hundred principles of management | |
| Drucker (1954): principle of organizational structure—centralized and decentralized structure-classification-federal, and functional | |

**Table 8.2** Streams of management thought-human relations and behavioral sciences

| Human relations | Behavioral sciences |
|---|---|
| Mayo (1945): problems of western civilization | MWeber (1900): proposed a theory called theory of authority structures. He described an ideal type of organization which he called Bureaucracy |
| Roethlisberger (1939): wrote the book management and worker, he speaks about the importance of social and psychological factors and how they play a much greater role than monetary incentives alone | Likert (1961): he worked on interaction-influence system that uses concept of supportive relationship between members of an organization as a central theme |
| Whyto (1955): in his book "money and motivation," he says social factors are more important in job performance than economic incentives. In his book, "The organizational man" he speaks about other directedness. Dominance of others and role playing based on external influence of other role senders | Moreno (1947): proposed "SOCIO-METRIC MODELS" which is used by organizations to create cohesiveness and high-performing teams |
| McGregor (1960): worked on hierarchy of needs developed by "Maslow" and was applying those concepts to industrial settings. Formulated Theory-X and Theory-Y and published "Human side of enterprise" | Viteles (1953): he defines morale in his Book "Motivation in industry's willingness to strive for goals by a particular group" |
| | Lewin (1951): proposed three-step models for managing organization change and development. Three steps are (a) UNFREEZING the status-quo (b) MOVEMENT to new state (c) REFREEZING the change to make it permanent |
| | Bavelas (1948): proposed the theory of COMMUNICATIONS NETWORK. The name of his paper is "A MATHEMATICAL MODEL FOR GROUP STRUCTURES" |
| | Chris Argyris (1968): integrating the individual and organization |
| | Selznick (1948): being a sociologist, he was interested in study of group and he proposed that organization is a dynamic system constantly changing and adapting to internal and external pressure and is in process of evolution |

**Table 8.3** Streams of management thought–quantitative approaches and managerial economics and accounting some people/task matrix approaches

| Quantitative approaches | Management economics and accounting |
|---|---|
| Shewhatt (1924): used statistical inference and probability theory in sampling inspection and in quality control by statistical means | A Marshall (1850): principles of economics |
| Von Neumann and Morgenstern—timing and pricing in a competitive market, military strategy (Game Theory) | JMClark (1923): a pioneer volume in relating the economic view of cost to needs of business—"Studies in the economics of the overhead cost" |
| Feller contributed on probability theory and its applications in myriad fields, especially in business. He also contributed to Markov Process, Dynamics of operational systems, and queuing process | McKinsey (1924): directed interest toward the use of accounting in management in his "Managerial Accounting" |
| Savage—key figure in field of statistics. He has worked on Bayesian statistics and stochastic processes and contributed to the development of decision theory | Terborgh (1949–58): investment decisions. Book: "Dynamic equipment policy," "Business investment policy" |
| Dantzing—assignment of equipment and personnel, scheduling, input–output analysis, transportation routing, product mix, allocation processes, and linear programming | Dean (1951): replacement of equipment through failure and deterioration |
| Schfailaer placed emphasis on human behavior in decision making; viewed as an identifiable observable and measurable process; increased attention given to managerial psychology | Vatter (1954): Tailor-making cost data for specific users |

- Human Relations
- Behavioral Sciences
- Quantitative Approaches
- Managerial Economics and Accounting.

These different streams of management with their contributors and contributions are presented in Tables 8.1, 8.2, and 8.3.

## 8.11 Summary

We defined management and engineering management in this chapter. We looked at the overlap between engineering and management and the interdependence of the two. We also examined the various management levels and the responsibilities at each level. We understood the various roles of a manager and the influence of each role on the overall objectives of the company. We then briefly looked at the various management thoughts and their influence on our understanding of management.

## 8.12 Review Questions

1. How can engineering and management satisfy the several parts of the definition of a profession?
2. From the different definition of management given by several authors, what has turned out to be the definition of management in the present context?
3. Differentiate the roles of a front-line manager, middle-level manager, and top-level manager.
4. Discuss the skills required at different levels of management, as given by Katz.
5. How is Engineering Management different from Management?
6. Trace the origins and growth of engineering management.
7. Discuss the contribution of Taylor and Weber to management thought.
8. What do the Hawthorne experiments bring out?
9. Discuss the quantitative philosophies of management along with their promoters.

## References

1. The Engineering Team (1979) New York: Engineers Council for Professional Development (now Accreditation Board for Engineering and Technology)
2. Dalton E (1979) Mcfarland, management: foundations and practices, 5th edn. Macmillan Publishing Company, New York, pp 4–5

3. Katz RL (1974) Skills of an effective administrator. Harvard Bus Rev 52(5):90–112
4. Selected excerpts from Henry Mintzberg, The Nature of Managerial work, chap 4 Harper-Collins Publishers, Inc, 1973
5. Fayol Henri (1949) Administration Industrielle ET Generale, constance storrs, trans. Sir Isaac Pitman & sons Ltd., London
6. Babcock DL (1929) The engineer's contribution to modern life an address to the American Institute of Mining and Metallurgical Engineers on receiving their Saunders Mining Medal at their 1928 annual meeting (reprinted in Jackson DC Jr, Jones WP (eds) The Profession of engineering, Wiley, New York, pp 119–120)
7. Heilmeier GH (1994) Room for whom at the Top? Promoting Technical Literacy in the Executive Suite, THE BENT of Tau Beta pi, Spring
8. Breaking Ground, Engineering (London), 1:1, January 5, 1866, p 1

# Chapter 9
# Foundations of Technology Management

Learning objectives:

- Understand strategic planning and strategic management
- Understand vision, mission, goal, and purpose
- Understand strategic decision making and models of strategic decisions
- Understand strategic management process
- Understand the qualitative tools used for strategic planning
- Understand technological forecasting
- Understand invention, innovation, and entrepreneurship.

## 9.1 Introduction

As discussed in Chap. 8, engineering management involves a synthesis of engineering and management. The function planning and management has undergone a drastic change in the present context. It is now strategic planning and not just planning. Today's manager has to concentrate on strategic decision making. In this chapter, we discuss all these and technological forecasting. Also discussed are some essentials for managing technology like invention, innovation, and entrepreneurship.

## 9.2 Strategic Planning

Planning provides a method of identifying objectives and then designing a sequence of programs and activities to achieve them. Planning is to decide in advance what to do, how to do, and who should do it.

An organization requires developing effective strategies for achieving its mission. Strategic Planning is the organized process for selecting these strategies.

Anthony defines strategic planning as: the process of deciding on the objectives of the organization, on changes in these objectives, on the resources used to govern the acquisition, use, and disposition of these resources [1].

R. Srinivasan, *Strategic Business Decisions*,
DOI: 10.1007/978-81-322-1901-9_9, © Springer India 2014

## 9.3 Purpose and Mission

Definition of organizational purpose and mission is an important and difficult task for the top management. This is more due to the fact that aims of organizations are very broad, and there exists certain vagueness regarding specific goals. As Drucker says, "Defining what our business is always a difficult question and the right answer is anything but obvious" [2].

There is semantic ambiguity between the terms purpose and mission. Some authors like Steiner [3] have used the terms purpose, mission, and objectives interchangeably. We can look upon the basic purpose of the firm as those fundamental ends and business lines the firm wishes to pursue. The basic mission statement is found in the corporate charter of a firm, but management has to choose those statements to which the firm is committed. According to Steiner, the purpose of a company should include larger societal obligations. Fundamental aims are determined implicitly and explicitly by the top management. The mission of the organization expresses its underlying thrust. This can be stated at different levels of abstraction. Operation-wise mission identifies the market of the firm, products, and competition. Voltas, for example, defined its purpose or mission as "profit, growth and excellence." Oil and Natural Gas Commission (ONGC) [4] defines its mission as: "To stimulate, continue and accelerate efforts to develop and maximize the contribution of the energy sector to the economy of the country."

## 9.4 Objectives and Goals

Goals can be classified into three types: official, operative, and operational. Official goals are general aims, which are described in the Memorandum of Association/Charter/Annual Report.

**Official goals** serve the purpose of public relations value and help in legitimizing the organization in the business environment.
**Operative goals** provide an indication of what an organization is really attempting to do. Generally, these can be inferred from the actual operating policy. They can help focus attention, reduce uncertainty, and provide a choice of organizational design alternatives to choose from.
**Operational goals** refer to those used by supervisory personnel to supervise the performance of subordinates and thereby influence their behavior.

Objectives and goals in an organization provide the foundation for all managerial activities. They can be considered as ends or aims toward which all activities are directed. According to Brown and Moberg [5], objectives and goals serve the following functions in an organization. They are as follows:

- Aid in legitimizing the organization.
- Assist in identifying inter-organizational relationships.

- Serve in building a public relations value.
- Help attract support from different agencies and also attract right people to the organization.
- Facilitate image building with the different players in the business environment, such as suppliers, customers, policy makers, and the government.
- Make coordination of the multiple tasks easier.
- Help in resolving conflicts.

While we can consider the goal as an open-ended statement of what an organization wants to accomplish without quantification and time criteria, objectives can be considered as providing both. Goals can be considered as motivators in an organization. They can pose a challenge to its members and generate commitment from them. Some of the areas where an organization can establish goals and objectives are as follows:

- Efficiency (reduction in costs)
- Profitability (increase in net profits)
- Growth (increase in total assets, sales, etc.)
- Wealth for shareholders (dividends, stock market appreciation)
- Resource utilization (return on investment, return on equity)
- Brand reputation (of being considered a leading firm)
- Contribution to employees (job security, compensation)
- Societal contribution (taxes, community service)
- Leadership of market (market share)
- Leadership in technology (innovation, creativity).

## 9.5 Program Strategies

We can refer to purpose, mission, and objectives in an organization as master strategies. Program strategies are those action plans or methods that are used to achieve an organizational objective. For example, an organization's long-term sales growth rate may be 35 % per year. In order to achieve this, it may likely formulate a relevant program strategy referring to development of new products, improvement of existing product line, quality certification tags, etc.

The main components of program strategies are programs, budgets, and procedures.

Budget is a statement of an organization's progress in monetary terms. Budgets are used as tools for planning and control. They give the detailed cost breakup of each program. Organizations often demand a certain percentage as return on investment (ROI), referred to as hurdle rate, before approving a new program. The emphasis is on the new program adding to organization's profits and building shareholder value. In this sense, budget provides a detailed plan of the new strategy in action and specifies through a proforma financial statement the expected impact on the organization's financial future. For example, Hero Honda, a leading player in the two-wheeler segment, increased its marketing budget by more than 100 %, to Rs. 5 crores, for the financial year, i.e., 2000–2001. This was essentially to counter the promotion of Bajaj Auto, which continues to give stiff competition.

Procedures, also referred to as standard operating plans, are the sequential steps or methods, which describe in detail a particular task. They describe the various activities to be carried out in order to complete the company's program. The procedure followed by Delta Airlines to cut costs and employees is illustrated in Exhibit 9.1.

**Exhibit 9.1 Strategic Management at Delta Airlines**

Delta Airlines initiated an aggressive cost–cutting program in April 1994, called Leadership 7.5, in order to become profitable in highly competitive industry. Because of deregulation, new competitors like southwest Airlines were able to introduce low cost strategies to offer extremely cheap fares (and minimal service) to gain market share, resulting in half-filled flights for full service companics like Delta Airlines. Delta Airlines had not turned a profit since 1990 and chose to institute a turnaround strategy (a type of retrenchment corporate strategy) to achieve an objective of reducing annual expenses by $2.1 billion by June 1997 (and make a profit). To fulfill this strategy, for which the company had long prided itself: lifetime employment, high pay, lush in-flight services, and routes to every destination. The Leadership 7.5 program attempted to reduce the amount of money spent on each airplane seat from 9.6 cents in 1994 to 7.5 cents in 1997 per flight mile.

The company budgeted $400 million in savings from marketing, $300 million from layoffs, and $310 million from onboard services. To reduce the number of employees, technical experts in hydraulics, metalworking, and avionics and other trades and experts were asked to design cross-functional work teams. Marketing expenses were cut by instituting a cap on travel agent commissions and emphasizing sales to bigger accounts. In addition to layoffs, purchasing and food service procedures were changed.

The success of turnaround strategy can be evaluated by measuring the amount the firm was spending on each airplane seat per flight mile. Before the "Leadership 7.5" program was instituted in April 1994, the cost per seat was 9.76 cents. By the end of 1995, it was down to 8.4 cents. The program seemed to be working, but Delta needed to reach 7.5 cents by June 1997 to achieve the corporate objectives of reducing annual expenses by $2.1 billion, and make a profit.

*Source* Business Week, December 11, 1995, pp. 106–107.

Programs can be looked upon as a collection of activities to achieve a certain end of a specific purpose. They have a finite life, and they are very clearly defined, for example, increase in labor productivity by 10 % in a year.

Program strategies have to take into account the changing market scenarios. With the onset of globalization, almost all the players in a market have had to redefine their program strategies according to the changing market conditions. The major emphasis of most of the firms operating in the market is to offer quality products and reduce costs. This is illustrated by the example of Delta Airlines, which adopted various procedures to cut costs. To effect employee reduction, Delta Airlines asked its technical experts in metal working, avionics, etc., to form cross-functional teams. To reduce marketing expenses, the airlines fixed an upper limit on travel agent commission and laid emphasis on bigger account sales. Further, purchase and food service procedures were also changed.

## 9.6 Stimulus for Strategy

A strategy in an organization may be the result of a triggering event. Such probable events could be the following:

1. **A new CEO** He can provide a dynamic leadership, which can boost the spirits of the employees and motivate them to perform better. Basically, the Chief Executive is lifting the veil of complacency and forcing employees to ask the reasons for the organization's very existence.
2. **Outside intervention** This may be in the form of financial intervention by lending agencies deciding to stop further financial assistance or by asking repayment in full of the existing assistance.
3. **Impending ownership change** The threat of a takeover by another company in the market.
4. **Failing performance levels** When there is gap between the desired and projected performance, the organization will be put to hard thinking on strategies to bridge the gap.

Exhibit 9.2 presents the case of Iomega Corporation, where the new CEO stimulated a change in strategy.

**Exhibit 9.2 Iomega Corporation-Triggering Event**

Iomega Corporation is a successful manufacturer of computer storage devices. Its most popular line of products is the Zip drive, a book—sized, portable storage device that uses a kind of floppy disk with a capacity of 100 megabytes-equal to about 70 standard floppy disks. Earning \$8.5 million on \$326.2 million of sales in 1995, the company's stock price escalated from \$5 per share in 1995 to \$112 (after adjustment for a stock split) in April 1996.

Until Kim Edwards took over as Iomega's CEO in 1993, the company had been an unglamorous provider of niche computer storage products. Soon after he joined the company, Edwards asked his team to name some potential new markets for the company's products. After a long pause, one person said,—the Air Force really likes our Bernoulli Box. Thought Edwards, Geez, this isn't good. The Bernoulli Box was a powerful storage device, but it was so expensive and specialized that only a few buyers, such as the military, had any use for it—I realized the company had no clue that there was a mass-market out there, waiting for a fun product, commented Edwards. Soon Iomega's engineers developed a series of products to appeal to this mass market: the zip drive, the ditto tape backup system, and the Jaz removable hard drive, which holds one gigabyte of data-ten times as much as Zip.

*Source* Wall street Journal, June 17, 1996 p. B6.

## 9.7 Strategic Decision Making

Strategic management is characterized by its emphasis on strategic decision making. As an organization grows bigger and becomes complex with higher degree of uncertainty, decision making also becomes increasingly complicated and difficult. Strategic decisions have to deal essentially with the long-term future of the organization and have three important characteristics.

1. **Rare** Strategic decisions are not common and have no precedents.
2. **Consequential** Strategic decisions involve committing substantial resources of the company and hence a high degree of commitment from persons at all levels.
3. **Directive** Strategic decisions can serve as precedents for less important decisions and future actions of the organization [6].

### 9.7.1 *Mintzberg Model*

According to Mintzberg [7], the modes of strategic decision making are the following:

1. **Entrepreneurial mode** Formulation of strategy is done by a single person in this mode. The focus is on opportunities. Strategy is guided by the founder's vision and is characterized by bold decisions. In the Indian setup, we can cite the case of Wipro InfoTech as an example of this mode of strategy formulation.
2. **Adaptive mode** This mode of decision making is referred to as "muddling through." It is characterized by reactive solutions rather than a proactive search for new opportunities. We can again cite the example of Wipro InfoTech introducing the sale of customized personal computers in response to Dell computers entering the Indian market.

3. **Planning mode** This mode of decision making involves systematic information gathering for situational analysis, generating alternate strategies, and selection of appropriate strategy. As could be inferred, this mode includes both the proactive mode and the reactive solutions to current problems. For example, entry of MNCs into automotive markets in India has made the lead player Maruti Suzuki to come out with new models and discard/slow down production of non-moving and old models.

Sometimes, organizations may adopt a fourth mode called the "logical incrementalization" mode. This is a synthesis of all the three modes of strategic decision making listed above. Quinn [8] describes logical incrementalization as "An interactive process in which the organization probes the future, experiments and learns from a series of partial (incremental) commitments rather than through global formulation of total strategies."

As has been made out, strategic decisions involve an interface between an organization and its external environment. This is in contrast to operating decisions, which are more frequent. The effects of strategic decisions permeate throughout the organization at all levels for a considerable length of time, since it involves committing substantial amount of resources of the organization. Sometimes, this is referred to as a non-self-generating decision, implying that though strategic decisions may be few in number, the organization should always be aware of the need to make such decisions.

Hence, it is essential to put a mechanism in place to take strategic decisions. This is truer in the case of Indian organizations with the onset of liberalization.

Table 9.1 provides a comparison of the operating and strategic decisions. As could be seen from the table, the organization's short-term survival is affected by its operating and administrative efficiency. Its success or failure in the long term depends on right strategic decision making, i.e., upon doing the right things rather than upon doing the things right.

## *9.7.2 Management by Objectives*

Concept of management by objectives (MBO) was formulated by Drucker [9].

MBO can help translate broad organizational goals and objectives into specific individual objectives. It is usually employed between superior and subordinate at every level in an organization; the steps involved are the following:

- Superior and subordinate should have understanding of goals and objectives of the organization and that of the superior's group.
- Superior and subordinate meet to establish objectives for subordinates for next year, which are consistent with the group objectives; they should be attainable and quantifiable to the extent possible (reduced inventory holding by 10 %).

**Table 9.1** Comparison of operating and strategic decisions

| What how | | Strategic decisions | |
|---|---|---|---|
| | | Clear | Unclear |
| Operations | Effective | I<br>Clear strategy and effective operations have contributed to success in the past and will contribute to success in the future | II<br>Unclear strategy but effective operations have contributed to success in the past, but success in the future is doubtful |
| | Ineffective | III<br>Clear Strategy but ineffective operations have sometimes worked in the past in the short run, but increasing competition makes success doubtful in the future | IV<br>Unclear strategy and ineffective operations have meant failure in the past and will be so in the future |

- A subordinate identifies the specific resources required to achieve the objectives. Objectives agreed upon should also include developmental objectives which can strengthen the subordinates' capabilities.
- Subordinate proceeds in the ensuring periods to proceeds over the ensuing periods to achieve these objectives. If problems occur, there can be more meeting between subordinates and the objectives with mutual agreement can be modified.
- At the end of the period, superior and subordinate meets to evaluate success of subordinate in meeting the desired objectives. This should be a constructive exercise and not an excuse for blaming each other. This is also followed by a set of new objectives for the following period.

- **Advantages of MBO**
  - Greater commitment and satisfaction on subordinates' part
  - Enforced planning and prioritization of future activities and more rational method of performance evaluation.

- **Disadvantages of MBO**
  - Time and paper work involved
  - Misuse when superiors simply assign instead of negotiating objectives
  - Gamesmanship of subordinates to negotiate easy goals.

MBO will not be a success if it does not have the initiation of and continuing support of the higher management in the organizations.

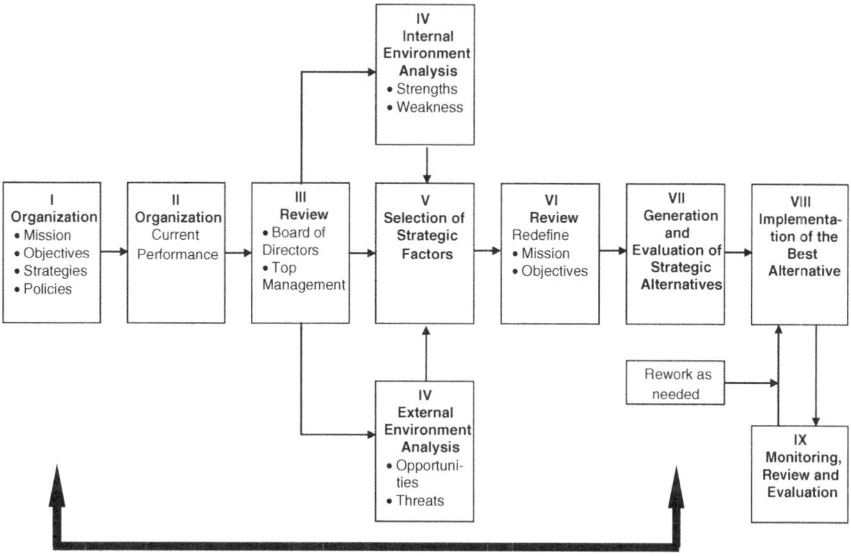


**Fig. 9.1** Strategic management process

## 9.8 Strategic Management Process

It would be better for an organization to follow the planning mode for the strategic management process. The planning mode, in addition to including the essential elements, is more rational and presents a better method of making strategic decisions.

Strategic management process that could be followed in a typical organization is presented in Fig. 9.1. The process takes place in the following stages:

1. The strategic planner has to define what is intended to be accomplished (not just desired). This will help in defining the objectives, strategies, and policies.
2. In light of stage 1, the results of the current performance of the organization are documented.
3. The Board of Directors and the top management will have to review the current performance of the organization.
4. In view of the review, the organization will have to scan the internal environment for the strengths, weaknesses, and the external environment for opportunities and threats.
5. The internal and external factors scan helps in selecting the strategic factors.
6. These have to be reviewed and redefined in relation to the mission and objectives.
7. At this stage, a set of strategic alternatives is generated.

8. The best strategic alternative is selected and implemented through program budgets and procedures.
9. Monitoring, evaluation, and review of the strategic alternative chosen is undertaken in this mode. This can provide a feedback on the changes in implementation if required.

An organization should stay current with technology to assure future profitability and make strategic planning for technology management an integral part of business strategy. Ericson et al. [10] opine that "The management of technology encompasses the management of Research, Product and process development and manufacturing engineering." According to them, there are three broad classes of technologies:

- **Base technologies**: These are a must for the firm for effective competition for a chosen product market mix. R and D should invest enough effort to maintain competence in these areas.
- **Key technologies**: These provide competitive advantage and enable differentiate the firm's product over that of competitor; it can also help attain greater efficiency, and in the production process, they should be given the highest priority.
- **Spacing technologies**: These can become tomorrow's key technologies. The critical issue is balancing support of key technologies to sustain competitive position and support of pacing technologies to create a vibrant future.

## 9.9 Forecasting

The most important premise in planning and decision making is the future levels of sales (for nonprofit activities—future operations). Everything depends on this—production levels which determine workforce size, training etc.; new facilities and equipment required; sales force size and advantage budget; level of inventories and accounts receivable. Forecasting comes in very useful here. Forecasting can be done by any of the following methods.

### *9.9.1 Qualitative Methods*

(a) **Jury of executive opinion**
In this method, executives of the firm make an estimate of future volumes and the head of firm does an average of these estimates. This method is economical. Another such method is Delphi method.

(b) **Sales force composite**
In this method, sales personnel prepare an estimate of sales in their own territory. The regional sales managers, according to how they judge individual

salesmen later on, adjust these estimates. Further additions to these figures are done by general managers to account for new products the company is willing to bring to market. Since sales people come close to the customer, this method is highly recommended.

(c) **Users' expectation**
In this method, Feedback is taken from customers to know their needs for the future. This is done through market surveys. This is an expensive method with the drawback being that customers more often do not have a clear idea of their future purchasing plans.

### 9.9.2 Quantitative Methods

(a) **Simple moving average**
This method is used when there is no trend in parameter over time. Forecast $F_{n+1}$ for the next period can simply be taken as the average of recent values, $A_t$

$$F_{n+1} = \frac{I}{n \sum A_t}$$

For example, if sales for years 2005, 2004, 2003, 2002, and 2001(n = 5) were 1,000, 2,000, 3,000, 4,000, and 5,000,
then, F 2006 = (1,000 + 2,000 + 3,000 + 4,000 + 5,000)/5 = 3,000.

(b) **Weighted moving average**
This method assigns weights to each year of the previous method. A recent year gets a higher weight for, e.g., year 2004 gets a higher weight compared to year 2003. The total set of weights $w_t$ should be equal to 1.0; for example, we are improving upon the previous method. Here,

$$F_{n+1} = \sum_{t=1}^{n} w_t A_t \text{ where } \sum_{t=1}^{n} w_t = 1.0$$

Using weights of 0.5, 0.4, 0.3, 0.2, and 0.1 in the previous example for the years 2005 to 2001 yields
$F_{2006}$ = 0.5(1,000) + 0.4
(2,000) + 0.3(3,000) + 0.2(2,000) + 0.1(1,000) = 2,800.

(c) **Exponential smoothing**
The previous methods have the disadvantage that it is cumbersome to remember the previous values and to give weights to each parameter being forecast. To overcome the disadvantages of the previous two methods, this

method reduces the weight of a value, as it gets older in a continuous manner. In this forecast for the next period, $F_{n+l}$ is given by the sum of

- the forecasted value of present period and
- some fraction $\alpha$ the difference between the actual $A_n$ and forecasted $F_n$ values for the present period.

#### 9.9.2.1 Regression Models

Regression models assume a linear relationship between the independent variable and dependent variables. They form a major class of explanatory forecasting models. In addition to providing useful forecasts, they also identify the causes and factors leading to the forecast value.

#### 9.9.2.2 Simple Regression Models

These models are based on the assumption that independent variable I depends on a single dependent variable D. Figure 9.2 gives a simple regression model, and Table 9.1 gives the calculation.

The regression problem is to identify a line

$$D = a + bI$$

such that the sum of squares of deviations between actual and estimated values is minimized. The two constants can be determined using the equations given below:

$$b = \frac{n\sum (I_i D_i) - \sum I_i \sum D_i}{n\sum (I_i)^2 - (\sum I_i)^2}$$

$$a = \sum \frac{D_i}{n} - b\sum \frac{I_i}{n}\bar{D} - b\bar{I}$$

where $\bar{D}$ and $\bar{I}$ are the mean values of $D$ and $I$, respectively, and $\Sigma$ indicates a summation from $i = 1$ to $n$.

We can calculate the trend line in Fig. 9.2 using the data in Table 9.2

$$b = \frac{4(8,500) - 6(5,200)}{4(14) - (6)^2} = 140$$

$$a = \frac{5,200}{4} - 140\left(\frac{6}{4}\right) = 1,300 - 140(1.5) = 1,090$$

and we can forecast a value for 2000.

$D_{2004}$ = 1,090 + (1,995–1,991)(140) = 1,090 + 560 = 1,650.

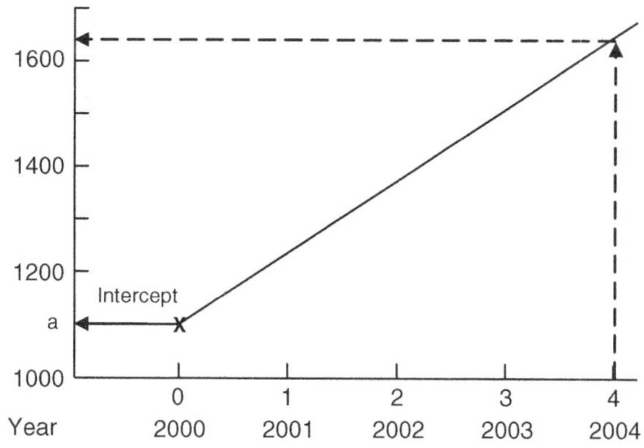


**Fig. 9.2** Simple regression model

**Table 9.2** Regression calculation data

| | $I$ | $D$ | $DI$ | $I^2$ |
|---|---|---|---|---|
| | 0 | 1,100 | 0 | 0 |
| | 1 | 1,300 | 1,300 | 1 |
| | 2 | 1,200 | 2,400 | 4 |
| | 3 | 1,600 | 4,800 | 9 |
| Σ | 6 | 5,200 | 8,500 | 14 |
| Mean | 1.5 | 1,300 | | |

Regression can be used to calculate the best straight line fit on a normal semi-log or log–log graph.

#### 9.9.2.3 Multiple Regression Model

This model is based on the assumption that the dependent variable $D$ depends on many independent variables $I$. Dependent variable can be proportional in many ways to the independent variables such as directly, inversely, etc.

## 9.10 Technological Forecasting

Today's world is changing at a very fast pace. Technology being the driver of such a change, it is necessary for engineers to have an idea of tomorrow's technology.

Shannon believes that [11]

1. Technological events and capabilities grow in a very organized manner;
2. Technology responds to needs, opportunities, and the provision of resources; and

3. New technology can be anticipated by understanding the process of innovation.

There are two types of technological forecasting.

1. Normative technological forecasting: In this method, some goal is made for the future and steps are chosen in a backward direction from the future to the present, which might enable to reach that goal.
2. Exploratory technological forecast: In this method, forecast is made from the current state of technology after considering the expected rate of technological progress. In this method, views of panels of experts are sought using Delphi method. Wide ranges of questions are asked to these panel members about the technological advancement in the future.

## 9.11 Strategies for Managing Technology

Managing technology requires skill sets that are different from the normal day-to-day management. We define two aspects of technology management that are different from, say, sales or operations management.

### *9.11.1 Invention and Innovation*

Inventors are people who come up with an idea for a novel product or process. But this is not enough. Take for example the development of Xerox copier. First conceived by Chester Carlson, a carbon chemist to copy legal documents in 1935, it was first neither economical nor efficient. However, it was issued patent in 1942 (by the USA). There were no takers for this invention till Batelle Memorial Institute, a nonprofit R and D organization, agreed to commercialize the product in return for share in royalties. Joseph Wilson, president of Haloid Corporation (a small company), took the risk of producing the first copiers based on Carlson's patents and Battelle's developments. The company grew and grew and changed its name to Xerox.

What Battelle and Haloid did constitute technological innovation: the introduction into the market place of new products, processes, and services based on technology [12]. Without innovation, invention creates little value; innovation should be an ongoing process: not just produce the first necessary product, should create a competitive product line. Product innovation and improvement forms a continuing part of technology strategy.

## 9.12 Entrepreneurship and Intrapreneurship

Entrepreneur is a person who organizes and manages a business undertaking assuming the risk for the sake of profit [13]. Today's organization requires continuous entrepreneurial activity to create new products and secure the future of the organization itself. This is referred to as Intrapreneurship.

## 9.13 Managing Technological Change

In order to manage technology, the organization needs to create a climate where intrapreneurs are encouraged to take risks, provided resources, and permitted initial failures, shifting to closer control of resources and costs as products/services become mature.

Top management in any organization must always keep abreast of changing technological trends. They must be willing to change. Business history is replete with examples of companies that failed to recognize in time new technologies that would replace their key products. Companies that were leaders in vacuum tube-based electronics find they are not wanted after products with large-scale integrated circuits (LSI) have entered the market. Companies in the present context should integrate technology strategy and business strategy to survive and prosper, for example, General Electric (GE), USA.

The Internet is a compelling example of how new technology can alter the way of business. Similar is the growth of cellular communication technology. Web pages allow companies to market to and service customers anywhere anytime, e.g., ITES operating in India servicing clients abroad, especially the USA.

## 9.14 Summary

In this chapter, we looked at strategic planning and strategic management process. In the overall scheme of things, we need this chapter to understand the influence of the organization's overall goal on project management. In this light, we need to understand what are goals, purpose, vision, mission of the organization, and their influence on the functioning of the company. We looked at the various strategic management models. Though representative, these give an indication of the thought process that goes into strategic planning. We have discussed the different methods of forecasting. These quantitative methods can be used to aid strategy making at the microlevel. We also looked at strategies for managing technology.

## 9.15 Review Questions

1. Why is planning important? What are the steps in planning process?
2. Consider companies in India whose purpose and mission has changed over history. Describe the change.
3. For a company of your choice in the Indian context, describe base, key, and pacing technologies that are important in strategic management of technology.
4. Discuss MBO and steps involved to implement it in an organization.
5. What are key result areas? Discuss w.r.t. an organization of your choice.
6. Discuss the different methods of technological forecasting using Indian examples.
7. For a given product and company of your choice, list the assumptions regarding economy, competition, labor, customer demand, etc., over the next 5 years.
8. Give an Indian example detailing how difficult it is to transform a technical idea into a successive product (innovation).
9. What are the personal characteristics of a successful entrepreneur? Take an examplc of your choice.

## References

1. Anthony RN (1965) Planning and control: a framework for analysis. Graduate school of Business Administration, Harvard University, Boston, p 16
2. Drucker Peter (1974) Management : tasks, responsibilities, practices. Harper & Row, New York
3. Steiner GA (1969) Top Management. Mcmillan, New York, p 32
4. Oil and Natural Gas Commission (ONGC) (1981–1982) Company annual report, New Delhi
5. Brown WB, Moberg DJ (1980) Organisational theory and management: a macro approach. Wiley, New York, p 234
6. Hiickson DJ, Butler RJ, Cray D, Mallory GR, Wilson DC (1986) Top decisions: strategic decision-making in organisations. Jossey-Bass, San Francisco, pp 26–42
7. Mintzberg H (1973) Strategy-making in three modes. Calif Manage Rev 16(2):44–53
8. Quin JB (1980) Strategies for change: logical incrementalism. Irwin, Homewood, p 58
9. Drucker PF (2006) The practice of management. Harper Business, Reissue edition
10. Erickson TJ, Magee JF, Roussel PA, Saad KN (1991) Managing technology as business strategy. Engineering management review, Spring, pp 34–38
11. Shannon RE (1980) Engineering management. Wiley, New York, p 43
12. Betz F (1987) Managing technology. Prentice-Hall Englewood Cliffs, New Jersey, p 6
13. (1986) Webster's new world dictionary of the American language 2nd college ed. Prentice Hall-Englewood Cliffs, New Jersey

# Chapter 10
# IT and Strategy

Learning Objectives:

- Understand the need to integrate IT and Strategy
- Understand the influence of IT on various aspects of Strategy.

## 10.1 Introduction

It would be difficult to deny that there is, or should be, a link between corporate strategy and information technology (IT) activities. Having said this, the statement is almost a truism which could be applied to any facet of business. One might similarly argue for the need to recognize the link between corporate strategy and advertising policy. For instance, an advertising campaign promoting a product line soon to be divested would clearly be a waste. In this chapter, we look at IT that can act as a facilitator for taking strategic decisions.

## 10.2 Strategic IT Investment

The problems with the processes involved in strategic IT investments are illustrated in Fig. 10.1. The rational, established view of the mechanism is shown in the outer spiral. Organizations appear to be prone to follow the inner spiral, however, due to circumvention of the process. This vicious spiral will lead to suboptimal decisions being taken and may be difficult to break out of.

The author wishes to acknowledge PHI Learning Private Limited for permitting use of material in this chapter from their publication, *Strategic Management—The Indian Context* (4th edition, 2012), authored by him.

R. Srinivasan, *Strategic Business Decisions*,
DOI: 10.1007/978-81-322-1901-9_10, 

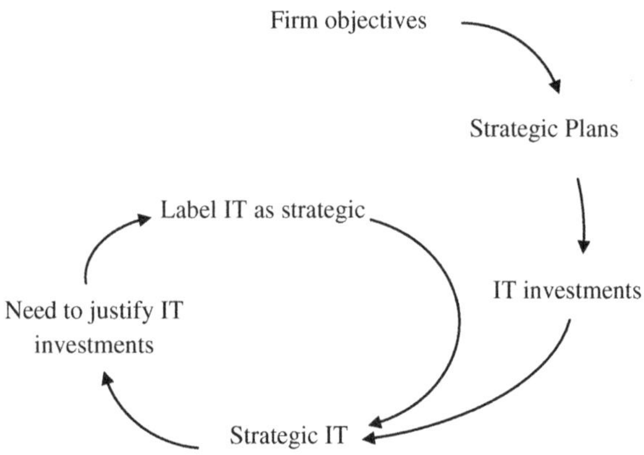


**Fig. 10.1** The vicious cycle of IT investment

Table 10.1 provides a summary of the traditional and alternative view of IT. Four dimensions are presented: nature, evaluation, use, and returns. As can be seen, there is a growing feeling in organizations, as per the alternate view, that the returns on IT investment may not be at the expected levels making the investment decision look suboptimal. This may also be due to the fact since IT is viewed as strategic, the usual evaluation process is not fully followed.

## 10.3 IT Strategy Components

Generally, sound strategic plans contain at least six components or elements. They include the following:

- Application System Components
- Application Development Components
- Infrastructure Component
- Maintenance Component
- Operations Component
- Security Component.

## 10.4 Viewing IT as Strategy

There are undoubtedly strategic aspects and uses of IT, but clearly not all use is strategic. Strategic aspects of IT suits more in terms of their influence on business results, such as changes in market share, improved product quality, increased market penetration, higher profit margins, and enhanced customer service. The top management feels the need to be convinced of the strategic impacts of information systems. Practising managers observe that strategic IT is supporting the

**Table 10.1** Traditional and alternative views of IT

| | Traditional view Strategic IT | Alternative view Strategic IT |
|---|---|---|
| Nature | Strategic IT investment flows from firm strategy | All IT investments are strategic |
| | Necessary investments are made | Labeling IT as strategic enables investment |
| | Strategic IT is an organizational benefit | Strategic IT is an organizational constraint |
| Evaluation | Evaluation of strategic actions is complex and not quantifiable | Evaluation of IT is complex and not quantifiable |
| | Projects are evaluated by quantification | Senior management preference for qualitative information |
| | Quantification dominates | Organizational politics dominates |
| | Strategy can be commulated | Strategy must be flexible and is left vague |
| | Quantification is beneficial | Quantification does not lead to superior performance |
| | Strategic IT should be evaluated | Strategic IT cannot be evaluated |
| | Evaluation used to be necessary | IT is moving into new areas |
| | Senior management dominates decision process | Technologists dominate decision process |
| | IT is just another investment | IT as infrastructure |
| Use | IT as technology | IT as a change agent |
| | IT to support structured activities | IT to support unstructured activities |
| | Use follows plan | Use conceived after implementation |
| | Technology follows strategy | Technology drives strategy |
| | Organizational participants know the strategy | Organizational participants do not know the strategy |
| | Strategic benefits come only from strategic systems | Strategic benefits can come from transaction systems |
| | System definition/use is in the hands of the developer | System definition/use is in the hands of the user |
| Returns | IT provides competitive advantage | IT, per se, gives only temporary benefits |
| | A champion is necessary | A champion is a hindrance |
| | Poor investment returns are to be avoided | Poor investment returns are inevitable |
| | Early adoption of IT is beneficial | Followers gain as well as leaders |
| | Post-audit should be carried out | Post-audit is a waste |
| | Success is good | Success may have negative consequences |

fundamental business goals and objectives; hence, investment should be in those IT-based applications and services that are likely to yield the best return in support of these. This, though, speaks of strategy as a preliminary screening device for investment considerations. Information systems for competitive advantage are those that reflect the fundamental objectives of the firm and that may have a significant impact on its success. The nature of this impact is via delivering lower cost production, differentiating the firm from its competitors, and identifying suitable market segments. These are achieved operationally by raising entry barriers, altering the balance of supplier/client power, creating dependencies for clients, offering new products, or changing the grounds of competition. If the information system performs any of these tasks, then it can be argued to be strategic. However, differentiation must be made between strategic information systems and those which, although flowing from corporate strategy, are designed to support the operational and tactical levels of the organization. The former systems support the strategic planning process while the latter enable implementation. Three characteristics of strategic information systems are being identified: link to be established among multiple parties, provide direct benefits to those parties, and affect competition that needs to be environment oriented. However, a resource should be considered as strategic when it is a significant portion of the investment base of the firm, and is not freely available on a competitive resource market. A concluding comment comes from a few researchers who turn the definition of strategic around, claiming that any system that confers significant competitive advantage can be regarded as strategic. From this definition flows the conclusion that what constitutes strategy may only be discovered ex post.

## 10.5 Influence of Information Technology on Pricing Strategies

Marketplaces in the twenty-first century contain various IT innovations, and the Internet has prompted many changes in business [1]. This research concentrates on three underlying factors that influence pricing:

1. increased availability of information
2. enhanced reach
3. expanding interactivity.

### *10.5.1 Increased Information Availability*

IT improves the gathering, handling, and analysis of pricing information. Availability of tremendous amounts of data (both online and offline; e.g., scanner data) enables firms to implement segment-based pricing, that is, determine an optimal price for each segment. With choice data, managers can microsegment a market on the basis of consumers' heterogeneity relative to their reservation, reference, and expected prices, as well as their price sensitivity and brand utility. Increased customer and competitor information also can increase price discrimination by enabling better demand forecasting. More sophisticated search tools give firms (rather than customers) better information about costs and prices. If used correctly, price transparency can favor those with greater access to search resources. The high-price companies such as Amazon.com and AOL pay for search bots. Technology advancements also can apply thousands of pricing algorithms in a matter of milliseconds to develop customized prices and discounts and provide a competitive edge to firms. Increased information gathering, handling, and analysis capabilities enhance price customization, bundling and unbundling, revenue management, and automated pricing strategies.

### 10.5.2 Enhanced Reach

Enhanced reach catalyzes various pricing strategies (e.g., auctions, revenue management, bundling/unbundling, price customization); in particular, the Internet provides companies access to an extended universe of customers, more demand, and new markets. Enhanced reach increases the popularity of online auctions that create a vast secondary market, unlimited by geography, time, or space. In this marketplace, not only do firms compete for consumer patronage but also consumers compete for a specific offering. Multiple customers from different geographic regions can participate in a single auction, and automated, intelligent agents enable single customers to monitor and participate simultaneously in several auctions. This larger pool of customers effectively increases demand, opens new markets, alters consumer behavior, and affects the valuation of products. According to some estimates, "Prices for hot (high demand) products are 17–45 % higher online than off because the web increases the likelihood of finding a buyer willing to pay a higher price."

### 10.5.3 Expanding Interactivity

IT may increase efficiency through electronic transactions and online customer interactions, which can affect pricing by creating exchanges (or online trading mechanisms), such as maintenance, repair, and operations hubs, through which buyers and sellers group together. Two models—matching buyers and suppliers and demand aggregation—result in a better price for buyers and more volume for suppliers, a win–win situation. Literatures discuss product valuation in online auctions in terms of opponent effects and quasi-endowments; the former involves the value of winning a bidding competition, whereas the latter refers to a sense of ownership/involvement that bidders develop during an auction. These dynamic effects can lead to behaviors that may differ from those of traditional purchase situations. As customers become more sophisticated in their use of search agents (e.g., shopping bots), the resulting ease of price comparisons may make buyers more deal prone.

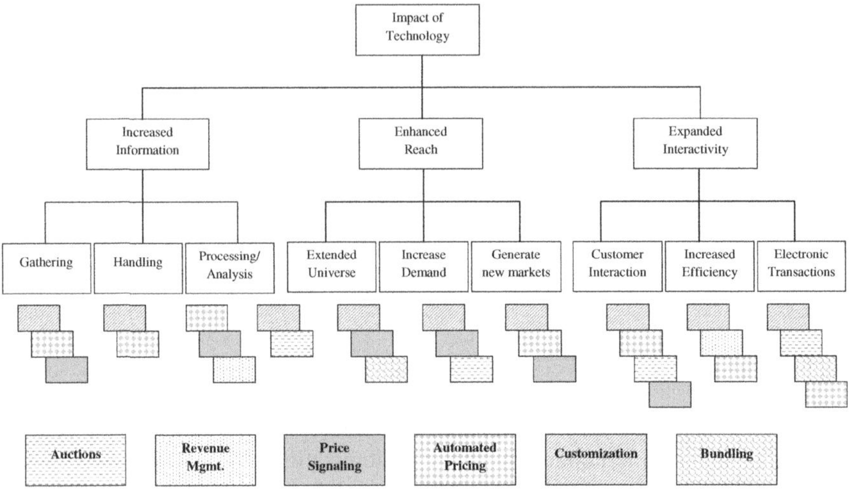


**Fig. 10.2** Influence of information technology on pricing strategies

Pricing managers can address this issue through various pricing mechanisms such as versioning, auctions, or group buying programs to increase the overall value they offer to customers. For example, revenue management may clear the market surplus so the firm can distribute excess capacity by offering different prices to different customer segments. Bundling products and services may reduce production and distribution costs. In summary (as depicted in Fig. 10.2), the combination of increased information, enhanced reach, and expanded interactivity prompts pricing strategies such as auctions, bundling, revenue management, signaling, automated pricing, and price customization.

## 10.6 Why the Emphasis on Strategy for IT?

It has been explained earlier that IT undergoes a development inside every organization, from a means to improve the efficiency and effectiveness of the organization to a means to influence the strategic position of the company, when applied more intensively. The way in which management controls IT has changed simultaneously (Fig. 10.3). In the first stages of automation, efficiency is the primary goal and the attention of management is mainly focused on technology. In this stage, the automation professional is an "extraterrestrial" expert, who decides what is best for the organization. In later stages, the effective functioning of the organization becomes as important a goal as efficiency. The management then becomes conscious of the fact that, next to technology, the design and structure of the organization is a decisive factor. User participation, information planning, and

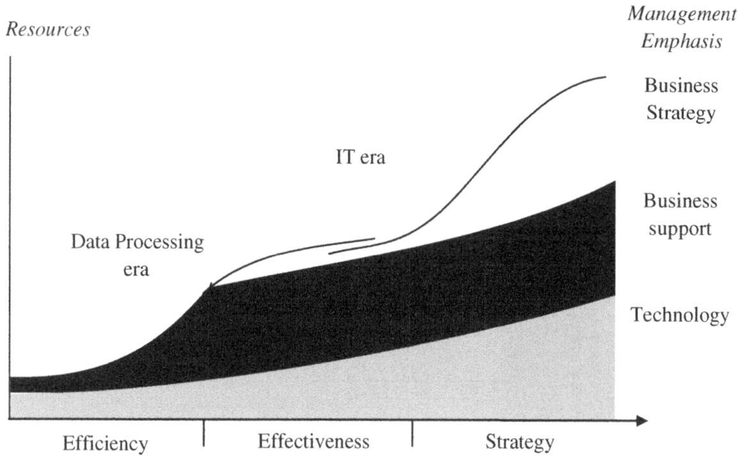


**Fig. 10.3** Nolan growth curve and the growth of management

the appointment of steering committees are manifestations of this. These organizations more and more recognize the need for a methodical approach to IT planning, as a result of disruptions in management, fusion or reorganization, cost increase, or new usage possibilities.

To an increasing extent, organizations that have more experience with automation realize that IT can not only improve the efficiency and effectiveness but also that it is of decisive importance to the company's success. IT usage planning then acquires a strategic quality in these organizations. IT functions as a catalyst in all this. These organizations set up business architecture and IT architecture, based on an objective, qualitative, and quantitative investigation into the current use of IT (Fig. 10.3).

## 10.7 What Can the Strategic Contribution of IT Comprise?

A good strategy cannot easily be copied by competitors because of the organizational, financial, social, and technical cost and the trial period involved in attaining the strategy. Every organization has its own profile, environment, and aims. Strategy indicates how the organizational structure should be designed and what the use of IT should be and should, therefore, be sufficiently concrete and specific.

A specific strategy leads to a unique interpretation of the architecture and the infrastructure. A good strategy focuses less and less on the product or the service itself and more and more on the rendering of services, reputation, and so forth.

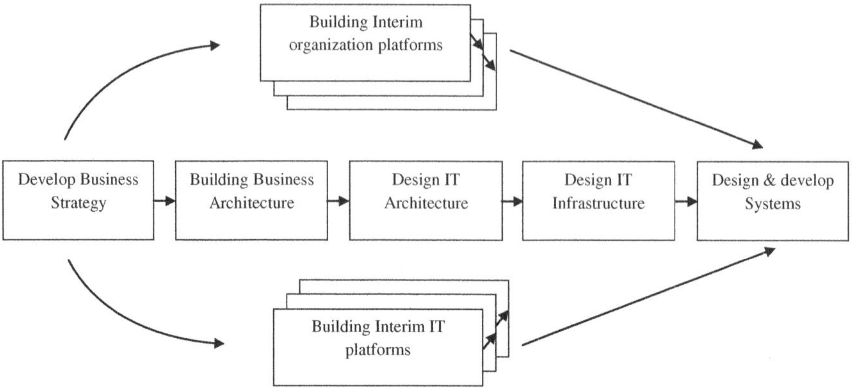


**Fig. 10.4** Architecture development process

This is especially applicable to products that have become a "commodity"; they are produced by many suppliers and their quality is usually high enough.

This is why strategies in the same branch differ to a high degree; for example, a great difference arises when the bottom of the market is the focus, for example, East European cars such as Skoda and Lada, or, on the other hand, the top of the market, for example, the German top cars, Mercedes and BMW. Quality control, tailor-made products, and product innovation are applied in a very different way. This leads to very different information needs and to a unique use of IT within the same car branch. Quality control, tailor-made products, and innovation are of vital importance, for example, to Mercedes. The use of IT should strongly support these business processes in particular and constantly offer the management of this car manufacturer the appropriate steering information described in a linear way, for the sake of clarity (Fig. 10.2). The development of an architecture starts with the business strategy. On the basis of this, the organization structure and organization processes will be designed in the business architecture (see below) and the "IT architecture" derived from this. The IT infrastructure is the technical filling-in of the IT architecture. "Implementation" deals with the introduction of the business architecture and IT architecture (Fig. 10.4).

## 10.8 Summary

We looked at strategic IT investments and their needs. We then looked at how IT influences various aspects of the strategy such as pricing and at how IT aids in improving the performance of the company by increasing the effectiveness and efficiency of a company.

## 10.9 Issues for Discussion

1. What is the theory behind strategic IT? What are its implications for marketing?
2. Discuss the scope of applying information strategy and IT for three industrial products of your choice in the Indian context.
3. For an international marketing firm, to what extent would you advocate the use of IT and strategy?

## Reference

1. Preston J, Hayward TE (1998) An assessment of the relationship between marketing, information and strategy formulation in the UK retail banking sector. 8(4):277–285

# Chapter 11
# R&D and Strategy

Learning Objectives:

- Understand the importance of R&D in a competitive environment
- Understand the R&D development process
- Understand the impact of R&D on strategy formulation and execution.

## 11.1 Introduction

The importance of technology as a competitive variable in the corporate strategy of industrial goods companies is widely recognized by business managers. This is especially true for high-technology industries where the performance of R&D often plays a crucial role in the success or failure of individual companies. The need to have an explicit R&D or technology strategy is therefore increasingly accepted among progressive managers. Such a strategy may consist of a large number of individual elements. These strategy elements have been classified into four main groups: the nature of products developed, the nature of markets sought, the nature of technology employed, and the orientation and nature of the new product process.

One important aspect of the R&D strategy, which is not given sufficient attention when analyzing the company's situation in terms of traditional market concepts, is the extent to which the company should cooperate with other companies or other external organizations in R&D. If the analysis is performed in terms of network concepts, the cooperation aspects become a central issue. Each company (or business unit) is then regarded as a member of a network consisting of companies and other organizations, which are linked to each other through

The author wishes to acknowledge PHI Learning Private Limited for permitting use of material in this chapter from their publication, *Strategic Management—The Indian Context* (4th edition, 2012), authored by him.

R. Srinivasan, *Strategic Business Decisions*,
DOI: 10.1007/978-81-322-1901-9_11, © Springer India 2014

various exchange activities. One of the major forces binding the network together is technological ties or bonds of different kinds. Thus, R&D has a direct impact on the company's relationships with other companies. As an example, a product development project may result in the need to change the interaction pattern within existing customer relationships or to establish new relationships with new types of customers, or with new types of suppliers or producers of complementary products. Close relationships with other companies or organizations may also be seen as potentially useful external R&D resources, which can be used in various ways in order to achieve internal technological goals.

The development of new unique products in close cooperation with technologically leading customers (or suppliers) can be an effective way to build competitive strength. Correspondingly, the lack of relationships with certain important units in the network may be an obstacle to technical development sources as well as the company's ability to control the technical development. A strategic question is how the company's own R&D resources should be used in relation to external units such as, for example, customers, suppliers, and different types of research and consultancy units which carry out related development work. There is also a geographical dimension to the cooperation strategy. The technology manager in companies operating internationally should, therefore, ask the question whether the firm should develop the same type of technical exchange relationships with foreign, as with domestic partners. The chosen strategy should then be taken into account when allocating and organizing the R&D resources within the company.

## 11.2 R&D Development Within Industrial Network

Empirical studies of industrial marketing and purchasing have given reasons for viewing the individual company more as an integrated part of a network than as a free and independent unit in a market. Each company has important relationships with other companies. These can go forward to distributors, customers, and end-users, backward to suppliers and sub-suppliers, or sideways to producers of complementary and competing products and to governmental or other types of public utilities (e.g., research institutes). The companies and other organizations are in this way tied together and form sets on different levels of the production and distribution chain. Individual companies are often members of several sets at the same time. On the input side, they can belong to different raw materials sets (such as the iron ore set, the wood set, the super alloys set, etc.) or technology sets (such as the hydraulic set, the electronics components set, etc.). On the output side, they can belong to different product-application sets, comprising the firms that are engaged as either manufacturers or consumers in the use of a particular product in a certain field of application (e.g., the use of microcomputers in industrial automation). Different sets are in turn linked with and dependent on each other, and constitute part of the total industrial network. The relationships are building up of diverse bonds between the companies. Six main types of bonds can be distinguished: technical, knowledge-based, temporal, social, economic, and legal. From the point of view of technical development, the technical and knowledge-based bonds are important. These bonds are created by the companies adjusting to each other in different ways.

## 11.3 Characteristics of the R&D Development Process

Technical development within an industrial network is a process of change in which new products and/or new manufacturing processes are devised and brought into use. The changes occur partly continuously, in the form of a large number of minor incremental steps and partly by leaps in the shape of major innovations. R&D activities that are dependent on each other occur simultaneously in different parts of the network. Some of these activities may take place mainly within a given unit. It may be a company that internally develops a new product, which is subsequently introduced in the market, or a new material, component, or process needed in its own manufacturing operations. In other cases, the development activities may be pursued jointly by several units in the network. It may be a selling and a buying company that jointly develop a new product or process, or a cooperation between companies on the same level of the production chain. Multiunit development, which takes place within established relationships, is common in many sets. This development within the company must be in accord with the development of the whole set. In other situations, companies interact voluntarily and opportunistically with other units in order to exploit potential synergistic effects. The companies seek a fruitful combination of internal R&D resources with external supplementary ones.

From an individual company's point of view, external units can be used in different ways. Customers may be used as sources of information and product ideas, as partners in laboratory or shop-floor tests of new product and as links to other important units in the network such as end-users. By developing close contacts with universities and research institutes, the company can get across to unique competences and laboratory equipment, which it cannot afford to acquire itself. Technical development within the industrial network is to a great extent a continuous process in which different technical parameters are successively changed and where several units play an active role in many and various modes of cooperation.

The technical development process in these cases is often characterized by a far-reaching interaction among different units. The interaction is sometimes forced and sometimes voluntary. For example, a company may be obliged to cooperate in product development with its customers in the knowledge that, in order to be accepted, the products must be custom-made.

## 11.4 Important Issues for the Individual Company

There are two interrelated issues concerned with R&D which arise when the company is considered part of a network. The first pertains to efficiency in the use of the company's technical and human resources. Because of the technical and knowledge-based bonds, certain of the company's own resources are invested in the relationships with other units in the network. Therefore, the R&D efficiency of a firm is not only a function of how its own assets are used internally, but also very much a question of how they can be used in order to activate and control external resources. The need to relate the company's own R&D resources to those of the network or individual sets, as we have already pointed out, pertains to both necessity and opportunity. The company has to adapt its development to that of the set or else influence the set to follow suit.

There are many possibilities for creating different forms of "cross-fertilization" by combining the company's own resources with those of others. These possibilities increase with higher degrees of specialization. An important question, then, is how the company should specialize in R&D in order to hold the best possible position, that is, the most efficient position, within the network. The second key issue for the individual company is its control over the technical development process. A company can enhance the efficiency of its own resources by cooperating with other units within the network, but it will at the same time lose some of its control over the process. A company can only exert complete control over its own R&D resources by isolating itself from the network. But this will also mean no control at all over the R&D resources of others. In order to obtain a modicum of control over the development of surrounding units, the company must develop close relationships with these. This requires investment of R&D resources in the relationships, which in turn leads to a certain control over the R&D resources. The issues facing the company can be seen as the establishment of a certain balance between its own R&D resources and those of external units, such as customers, suppliers, or research units. This balance will in turn determine both the efficiency of its use of its own R&D resources and the control of them and of external R&D resources.

## 11.5 Special Issues for International Companies

Internationally operating companies are at the same time members of several national or regional sets. These geographically separated sets often exhibit differences with regard to technical matters. The level of industrialization may vary among the company's different markets. Even within the group of the most advanced countries, there can be great differences. Because of the industrial structure and earlier development pattern, certain countries hold pride of place within certain product areas while others dominate the technological development elsewhere. Divergent local traditions are frequently manifested in the sets, which mean that product requirements vary or that different types of technical solutions are predominant in different countries. Furthermore, large individual customers in important national sets may have special needs, which lead to demands for local adaptation of products or production. All these circumstances create special problems with regard to technical development, but also special opportunities for international companies compared with those that operate only in a national or local set.

## 11.6 Location of R&D

A first and very important issue concerns the geographic location of the R&D resources. R&D is highly centralized in the home countries of multinationals. Conference Board in the USA shows that only 10 % of American companies' expenditure on R&D occurred outside the USA during the period 1966–1975. There is evidence that other multinationals, with the possible exception of Canadian firms, behave as the Americans in respect to R&D location. Another study showed that 86 % of all R&D done by Swedish companies was carded out in Sweden. It is also discovered that R&D operations were less centralized the longer the firm had been engaged in international business and the larger this business was in relation to its total turnover. This is clearly in line with an evolutionary pattern of foreign R&D investments.

An extensive investigation of the R&D investments was made by seven U.S.-based multinational organizations. According to these findings, the vast majority of foreign R&D investments seemed to follow a pattern where small investments are first made in technical service laboratories in order to help transfer U.S. technology. Later, these R&D investments usually expand and evolve into organizational units seeking to develop new and improved products and processes expressly for foreign markets. Given the proper preconditions, these R&D investments may expand further and evolve into "global technology units," developing products and processes for simultaneous manufacture in several major world markets.

## 11.7 New Opportunities

Being an international company operating in a large number of national sets can also confer certain advantages from the point of view of technical development. New development opportunities can be created by capitalizing on differences between countries. By combining technologies originating from different countries, unique products or systems can be developed. For example, the high Swedish competence within the office automation field in combination with US computer technology should offer interesting development opportunities to companies like Ericsson, which have R&D resources in both countries. Furthermore, the fact that a company operates internationally can make it more attractive as a partner in development projects. These problems and opportunities give rise to important strategic and organizational issues with regard to the technical development work in international firms. First, how should the R&D work be pursued in relation to external units in the different national sets? And second, given a certain cooperation strategy, how should the R&D resources be organized both functionally and geographically? The first of these issues will be discussed in the next section, where different cooperation strategies will be identified first for national and then for international companies. The classification of strategies then leads to the subsequent section to proposals for the geographic organization of R&D resources.

## 11.8 Cooperation Strategies

### *11.8.1 National Companies*

The R&D activities of national companies can be characterized based on the degree to which they are conducted in cooperation with external units. At one extreme of this dimension, all activities are carded out within the company: The company can then be said to follow an Introvert R&D strategy. This means that new products or processes are based on internal discoveries or knowledge about market needs in general and developed without any external cooperation. At the other extreme, all main activities are carded out in cooperation with external units. In this case, the company is said to follow an Extrovert R&D strategy. Several forms of cooperation strategies can be identified on the scale between the two extremes. For example:

1. Coordination of the internal R&D activities with other firms through extensive exchange of ideas and information. The technical development is constantly discussed with customers and/or suppliers. The internal R&D activities are thereby oriented toward concrete customer problems and may lead to the development of new products which the customers will need in the near future. The external units are in this case used primarily as sources of information and ideas, while the development work itself is mainly carried out within the company.
2. Joint testing of new products under development. In this case, field testing is carried out in cooperation with certain customers and/or suppliers at an early stage of product development projects (before the final specifications are set). The effects are often far-reaching adaptations to current customer needs and/or suppliers' abilities.
3. Joint development of new products or processes. Researchers and other technical personnel from the company and its suppliers and/or customers work together on certain projects. In certain cases, a joint project organization is created, based on a formal contract. In other cases, the R&D activities are less coordinated.

The company may have different cooperation strategies for different segments of its operations and in relation to different sets, e.g., vis-à-vis customers (product development), suppliers (product, component, or process development), and parallel units such as competitors or producers of complementary products.

### *11.8.2 International Companies*

When a national company pursuing an introvert R&D strategy goes international and starts marketing its products in several countries, there is no special problem with regard to R&D provided that buyers in different countries have similar needs and wants. This is often the case, for example, for companies selling equipment to scientists. Because of the rapid diffusion of information within the scientific community, the same scientific methods are commonly used in research laboratories throughout the world. For this reason it is frequently possible to sell identical products, such as instruments and accessories, in many different markets. For national companies pursuing an extrovert R&D strategy, the step to internationalization is more problematic in this respect. It is possible to distinguish between three alternative strategies that may be used by international companies wishing to carry on their R&D activities in cooperation with external units. The company can cooperate with:

- Partners in several different countries (including the home country).
- Domestic partners only.
- Only foreign partners in a few leading countries.

In the remainder of this section we shall discuss these strategies in greater detail. A strategic question for an internationally operating company is whether it should develop the same type of relationships with foreign, as with domestic business partners. That is, should it pursue an extrovert strategy in several different countries? Such a strategy will tax the ability of the company to handle an extensive technical exchange with a large number of distant units, although it offers potentially more opportunities.

The geographical location of the R&D resources and the division of responsibility between the R&D and marketing functions with regard to the external technical exchange become important organizational issues. If the company chooses to centralize the R&D resources fully in the parent company, the information gap between the R&D unit and the foreign partners becomes a problem. The researchers must devote considerable time to traveling around, or a great proportion of the information flow between R&D and the external units may have to be handled by an internationally deployed marketing organization. In the latter case, the sales personnel must possess sufficient technical competence to understand various technical matters and be able to communicate with the technical specialists in the external units as well as in their own company.

### 11.8.3 Domestic Partners

An alternative and less resource demanding strategy for a company that has the ambition to integrate R&D with external units is to limit the development cooperation to domestic partners (provided that there are any available). For example, such a domestically directed extrovert R&D strategy can mean that new products are developed together with domestic customers and subsequently exported to foreign markets. A goodly number of Swedish industrial goods companies seem to have followed this type of strategy (consciously or unconsciously).

### 11.8.4 International Partners

When there are no competent domestic partners available with whom the company can develop new products for the international markets, it may have to establish cooperation with foreign partners, i.e., choose an internationally directed extrovert R&D strategy. If the company's competitiveness is highly dependent on foreign technology (on the input or output side), it may even be preferable to locate all or part of the R&D operations overseas.

It is not unusual under these circumstances that the applications-oriented type of development activity is decentralized to those subsidiaries that have the most demanding and qualified customers. Proximity to suppliers of key components and raw materials can be another reason to establish foreign R&D units. One Swedish manufacturer of laboratory equipment considered moving its product development operations to the USA since many of the leading suppliers of electronic components are situated there.

Probably, product development is easier to internationalize than process development, the reason being that the latter should normally be carded out close to the production units (unless it is a matter of developing a radically new process). It is possible to cooperate with domestic partners in product design and foreign partners in component or process development, and vice versa. Thus, there are several possible cooperation possibilities, as outlined in Table 11.1. The four main cooperation alternatives are as follows:

1. Introvert R&D (i.e., no technology cooperation)
2. Extrovert R&D, with domestic partners only
3. Extrovert R&D, with foreign partners only
4. Extrovert R&D, with domestic and foreign partners.

**Table 11.1** Alternatives for technology cooperation

| Market orientation | R&D organization | Technology strategy | | | |
|---|---|---|---|---|---|
| | | Introvert | Extrovert | | |
| National | Centralized | | Only domestic partners | Only foreign partners | Domestic and foreign partners |
| International | Centralized | | | | |
| | Decentralized | | | | |

## 11.9 Patterns of R&D Strategy

Of course, like the national company, the international company can apply different strategies for different product groups or business areas and, as mentioned above, for different types of R&D activities. It is also reasonable to believe that the R&D strategy evolves in parallel with the internationalization of a company's sales. It is noted that the larger markets have the earliest and largest share of the multinationals' foreign activities. Thus, three countries, i.e., Canada, the UK, and West Germany account for almost two-thirds of the overseas R&D of US companies. European firms conduct above average amounts of R&D in the large American market. The suitability of the different strategies depends on the characteristics of the company and the set. The company can vary in such dimensions as total resources that are accessible, its production and organization structure, its technical competence, and its strategy. The set can vary with respect to the degree of concentration, homogeneity, geographic density, and stability of members of the set. Figure 11.1 shows the characteristics of the company and the set that influences the choice of cooperation strategies. In the following two sections, we shall discuss how two key characteristics, the concentration of R&D in the set and the production structure of the company, affect the suitability of the cooperation strategies for a company.

## 11.10 TechStrategy for R&D Planning

TechStrategy is a framework for technology road mapping, especially for the R&D sector, designed to support the needs of a specific program or project by providing a framework for planning and coordinating R&D efforts with operational requirements. Compared to the product technology roadmap, R&D-purposed TRM requires greater efforts in terms of selecting technology areas to be developed. Setting development targets in each technology area also requires more sophisticated techniques. Consequently, TechStrategy is a TRM technique that can give more weight to R&D planning activities by identifying technology needs, evaluating the current state of a firm's technology compared to its competitors', and

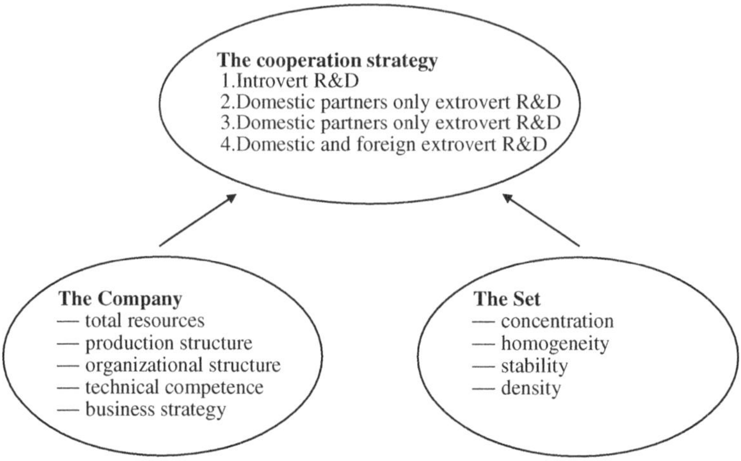


**Fig. 11.1** Factors influencing the value of different cooperation strategies

producing a development plan to deal with these needs and fill the technological gaps revealed by the process. This process consists of six consecutive phases. The first phase, *TRM Initiation*, consists of preliminary TRM development activities. Specifically, a TRM team is organized, TRM reports are designed, and a road mapping schedule is completed. The second phase, *Subject Selection*, is the starting point of real road mapping. At this stage, customer needs are gathered and items to be improved are proposed and selected. Where the TRM is developed at the firm level, the firm's operation divisions will be the operational sector for suggesting such items: where the development is at industry or government level, the operational sector will be at individual firm level. The third phase, *Technology Needs Assessment*, identifies specific technology needs for selected items. At this phase, each item is decomposed into its components to select what should be developed within the available resources.

If the item unit is a product, its components will be its parts, while if the item unit is a technology, the component is its sub-technology. The components most urgently in need of improvement are then selected, and in the fourth phase, *Technology Development Plan*, development goals, and strategies are created for each component. At the fifth phase, *TRM Implementation*, the TRM report, which details both the future outlook of the item (the macro TRM) and the development plan for its components (the micro TRM), is finalized and released for implementation. This information provides the basis for R&D budgets and project plan. The final phase is *Follow-up Activity*. Figure 11.2 shows the overall TechStrategy framework.

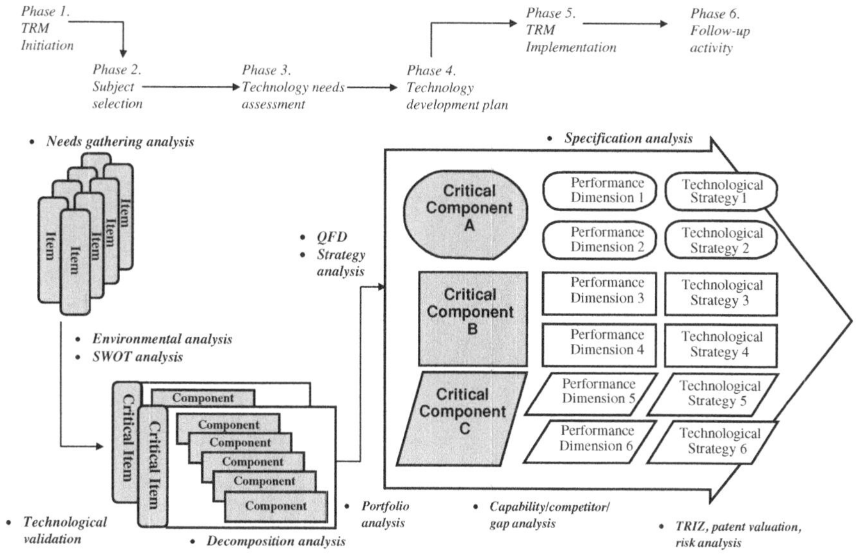


**Fig. 11.2** Overall framework tech strategy SWOT

# 11.11 Summary

In this chapter, we looked at how R&D aids improve the competitiveness of a company. We looked at the R&D development process and the factors that influence the R&D efficiency of a company. We then looked at the cooperation strategies that may be followed by companies and the situations in which partners may be chosen. We concluded the chapter by analyzing the various patterns in the R&D strategy.

# 11.12 Issues for Discussion

1. Take a company of your choice and develop a road map for its R&D planning.
2. Is R&D just a game of chance or can strategy make a difference?
3. What is the impact of global technology on industry's R&D strategies?
4. Explain the terms:

    a. Networking in R&D
    b. TechStrategy in R&D.

# Chapter 12
# Managing Projects

Learning Objectives:

- Define project management
- Understand how project management is not normal management
- Understanding the dependence of project management on strategic management
- Understand the steps in project management
- Understand the various project planning tools
- Understand what is a project-driven organization.

## 12.1 Introduction

In this chapter, we define a project, project management tools used in project planning, PERT, CPM and other variations, and resource allocation. We then discuss the elements of profit-driven organization. We discuss matrix organization structure, project manager and his role, motivation through team building, and conflict management. Also considered are the key project monitoring and controlling tools.

### *12.1.1 What is a Project?*

The Project Management Institute provides the following definition of a project: A project is a temporary endeavor undertaken to create a unique product, service, or result. A project is not the daily work to be completed by an employee. Projects are not routine, repetitive tasks! Ordinary daily work typically requires doing the same or similar work over and over, while a project is done only once. A program is a group of related projects designed to accomplish a common goal over an extended period of time. Each project within a program has a project manager. The major differences lie in scale and time span. Program management is the process of managing a group of ongoing, inter-dependent, related projects in a coordinated way to achieve strategic objectives.

R. Srinivasan, *Strategic Business Decisions*,
DOI: 10.1007/978-81-322-1901-9_12, © Springer India 2014

## 12.2 Project Characteristics

A project is a collection of tasks aimed toward a set of objectives. It has a definable end point and a finite life span. Project management technique will be particularly useful when the project calls for interaction between divisions, technologies, or organizations, and there are tight time and cost schedules to be adhered to. Project management has three essential considerations:

1. Time (project schedule)
2. Cost (monetary and non-monetary), and
3. Performance (the extent to which objectives are achieved).

Project management calls for difficult trade-off decisions.

The project proposal process involves proposal effort as soon as the Request For Proposal (REP) is received. Many organizations have marketing groups identifying the needs of a potential customer before the RFP is received, so that their response could be tailor-made to these needs.

The project proposal will have management, technical, and cost sections. A typical organizational chart for a major proposal is given in Fig. 12.1 [1].

In a sense, project managers perform the same functions as other managers. That is, they plan, schedule, motivate, and control. However, what makes them unique is that they manage temporary, non-repetitive activities to complete a fixed life project. Unlike functional managers, who take over existing operations, project managers create a project team and organization where none existed before. They must decide what and how things should be done instead of simply managing set processes. They must meet the challenges of each phase of the project life cycle and even oversee the dissolution of their operation when the project is completed.

## 12.3 Project Management: An Integrative Approach

Integration is designed to improve project management in the whole organization over the long haul. The rationale for integration of project management was to provide senior management with:

1. An overview of all project management activities
2. A big picture of how organizational resources are being used
3. An assessment of the risk their portfolio of projects represents
4. A rough metric for measuring the improvement of managing projects relative to others in the industry
5. Linkages of senior management with actual project execution management.

Operationally, what does project management integration mean? It necessitates combining all of the major dimensions of project management under one umbrella.

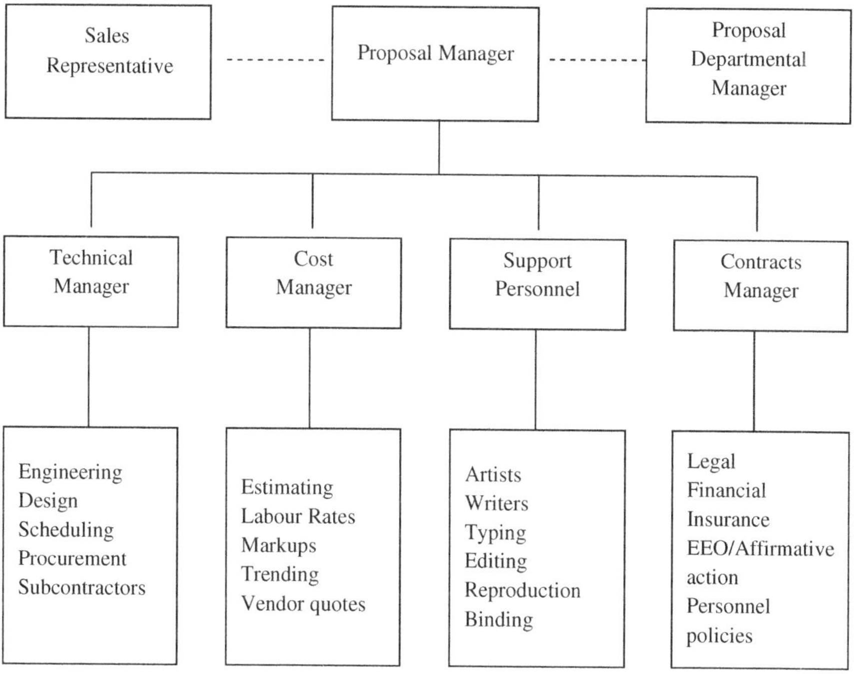


**Fig. 12.1** Typical organizational charts for a major proposal

Each dimension is connected in one seamless, integrated domain. Integration means applying a set of knowledge, skills, tools, and techniques to a collection of projects in order to move the organization toward its strategic goals.

We earlier had a chapter explaining strategy, strategic management, and the strategic management process. We now relate that chapter to project planning. It is very important that a project manager understands project management.

J. P. Descamps has observed that project managers who do not understand the role their project plays in accomplishing the strategy of their organization tend to make the following serious mistakes:

1. Focusing on problems or solutions that have low priority strategically.
2. Focusing on the immediate customer rather than the whole marketplace and value chain.
3. Overemphasizing technology as an end in and of itself, resulting in projects that wander off pursuing exotic technology that does not fit the strategy or customer need.
4. Trying to solve every customer issue with a product or service rather than focusing on the 20 % with 80 % of the value (Pareto's Law).
5. Engaging in a never-ending search for perfection that no one except the project team really cares about.

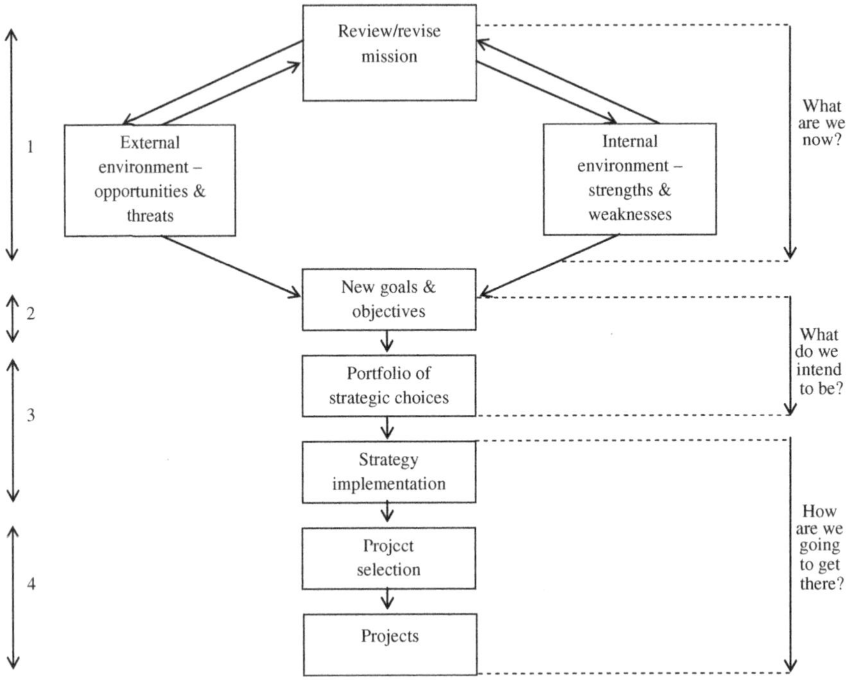


**Fig. 12.2** How to plan projects? (*Source* Project Management—Erik Larson, Clifford Gray)

## 12.4 Four Activities of the Strategic Management Process

The typical sequence of activities of the strategic management process is outlined here; a description of each activity then follows:

1. Review and define the organizational mission
2. Set long-range goals and objectives
3. Analyze and formulate strategies to reach objectives
4. Implement strategies through projects (Fig. 12.2).

## 12.5 How do We Plan Projects?

### *12.5.1 Project Scope*

Defining the project scope sets the stage for developing a project plan.

Project scope is a definition of the end result or mission of your project—a product or service for your client/customer. The primary purpose is to define as clearly as possible the deliverable(s) for the end user and to focus project plans. As fundamental and essential as scope definition appears, it is frequently overlooked by project leaders of well-managed, large corporations.

### 12.5.2 Establishing Priorities

Quality and the ultimate success of a project are traditionally defined as meeting and/or exceeding the expectations of the customer and/or upper management in terms of cost (budget), time (schedule), and performance (scope) of the project. One of the primary jobs of a project manager is to manage the trade-offs among time, cost, and performance. To do so, project managers must define and understand the nature of the priorities of the project.

### 12.5.3 Creating the Work Breakdown Structure (WBS)

Once the scope and deliverables have been identified, the work of the project can be successively subdivided into smaller and smaller work elements. The WBS is used to link the organizational units responsible for performing the work. In practice, the outcome of this process is the organization breakdown structure (OBS). The OBS depicts how the firm has organized to discharge work responsibility. In many cases, the size and scope of the project do not warrant an elaborate WBS or OBS. One tool that is widely used by project managers and task force leaders of small projects is the responsibility matrix (RM). Also called a linear responsibility chart, it summarizes the tasks to be accomplished and who is responsible for what on a project. In its simplest form, an RM consists of a chart listing all the project activities and participants responsible for each activity.

### 12.5.4 Project Communication Plan

Once the project deliverables and work are clearly identified, following up with an internal communication plan is vital.

An example of a communication plan is shown in Table 12.1.

### 12.5.5 Estimating Project Costs

Quality time and cost estimates are the bedrock of project control. Past experience is the best starting point for these estimates. The quality of estimates is influenced

**Table 12.1** Communication Plan

| What information | Target audience | When? | Method of communication | Provider |
|---|---|---|---|---|
| Milestone report | Senior management and project manager | Bimonthly | E-mail and hardcopy | Project office |
| Project status reports and agendas | Staff and customer | Weekly | E-mail and hardcopy | Project manager |
| Team status reports | Project manager and project office | Weekly | E-mail | Team recorder |
| Issues report | Staff and customer | Weekly | E-mail | Team recorder |
| Escalation reports | Staff and customer | When needed | Meeting and hardcopy | Project manager |
| Outsourcing performance | Staff and customer | Bimonthly | Meeting | Project manager |
| Accepted change requests | Project office, senior management, customer, staff, project management | Anytime | E-mail and hardcopy | Design department |
| Oversight gate decisions | Senior management and project manager | As required | E-mail meeting report | Oversight group or project office |

by other factors such as people, technology, and downtimes. The key for getting estimates that represent realistic average times and costs is to have an organization culture that allows errors in estimates without incriminations.

## 12.6 Project Planning Tools

There are a number of project planning tools that should be understood:

1. Statement of Work (SOW)
2. Milestone schedule
3. Work Breakdown Structure (WBS)
4. Gantt chart
5. Network Scheduling—PERT & CPM and Resource allocation methods.

Statement of work (SOW) describes exactly what is to be provided in the project. Milestones give key dates for major project activities such as preparation of design, first prototype, and first production item. This will help in detailed planning. Work breakdown structure gives the level-by-level subdivision of work to be performed in a project. It can help in:

- Describing the total project effort
- Plan and schedule effort
- Estimate costs and budgets
- Draw support network schedule
- Fix responsibilities and authorize work
- Monitor time, cost, and performance.

Gantt chart, developed by Henry L Gantt in 1915, is used in job shop or batch production environment.

Gantt charts require:

1. Tasks (activities) needed to complete the project
2. Precedence relationships of which tasks should be completed before other tasks can begin
3. Expected Duration of each task.

Table 12.2 explains this taking the example of a house project.

Figure 12.3 gives Gantt chart for the above project. Gantt chart is effective for small projects without an excessive number of tasks.

The two scheduling techniques PERT (Project Evaluation Review Techniques) and CPM (Critical path method) were developed in 1958 by Booz, Helen, and Hamilton (management consultants) and Lockheed Aircraft for use in ballistic missile development. Both these are discussed in detail in Chap. 5.

**Table 12.2** Planning information: Housing project

| Task | Follow task(s) | Weeks duration | Task description |
|---|---|---|---|
| A | – | 1.0 | Clear site |
| B | – | 1.0 | Obtain size stone and other basic materials |
| C | – | 2.0 | Obtain other basic materials and components |
| D | B | 2.0 | Erect pillars |
| E | B | 2.0 | Put beams |
| F | D, E | 2.0 | Erect walls |
| G | E, F | 2.0 | Complete roof |
| H | G | 2.0 | Finish interior |
| I | G | 6.0 | Finish exterior |
| J | I | 4.0 | Clean up site |
| K | H, J | 0.5 | Final inspection and approval |

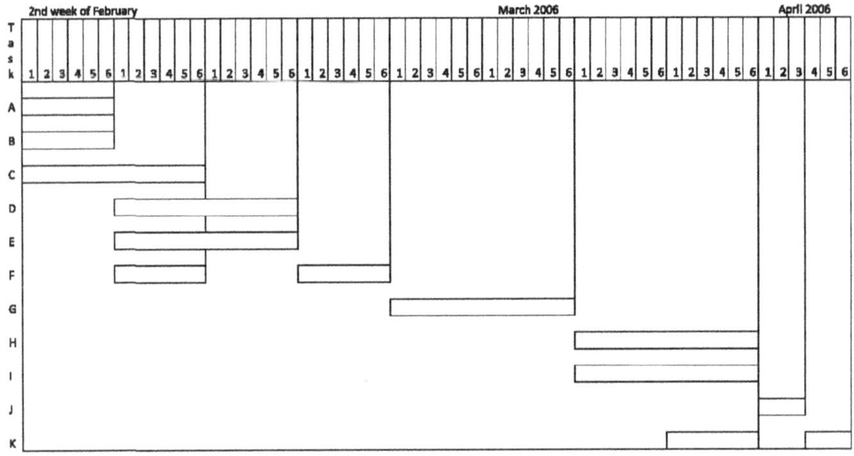


**Fig. 12.3** Gantt chart for simple housing project

Some of the well-known variations of PERT and CPM are:

## Precedence Diagramming Method (PDM)

This allows the flexibility of starting a succeeding activity even when the previous activity has not ended. This is made possible by inserting "finish to start" (FS) and "finish to finish"(FF) constraints—for example: In house construction, we may want curing for concrete for 2 weeks before building upon it. In PDM, we can connect the activities of *pouring concrete* and *cure concrete* with an FS delay constraint of 2 weeks.

**Graphical Evaluation and Review Technique (GERT)**

Whereas in PERT and CPM successor activities begin with certainty after predecessor activities are complete, in GERT it is assumed that alternate paths are possible, each with a probability—for example: Suppose we are testing a component which has a 0.9 probability of success and a 0.1 probability of failure, then GERT can estimate time durations by running them many times in Monte Carlo simulation.

## 12.7 Project-Driven Organization

In a large construction company like L&T or a space corporation like ISRO, work is carried out through projects. The sum of the profits earned through each project make up the total profits of the organization. In contrast, in a non-project-driven organization, e.g., small-scale industry engaged in manufacture of grinding machines, profit and loss is measured as vertical or functional lines.

Revenue producing functional line activities gets preference in resource allocation compared to projects. A project-driven organization has the following elements:

1. Project office—bearing responsibility for the project. It will have a project engineer responsible for technical, cost, and schedule of engineering activities. There will be a project engineer responsible for technical, cost, and schedule of engineering. There will also be project administrator responsible for project planning, control, and documentation.
2. Key support functions—essential for project success—can be systems, design and analysis, quality, production planning, installation and test, training, logistics, field support, etc.
3. Manufacturing and routine administration—includes all service activities such as accounting and finance, personnel relations, maintenance.
4. Future business—includes non-project-specific R&D and marketing as may be required.

## 12.8 Functional Organizational Structure

The functional (or traditional) organizational structure has survived for more than 200 years. It is functional since the organization is divided with functional areas such as marketing, finance, production, personnel, and R&D. A typical functional organization structure is shown in Fig. 12.4.

The merits/advantages and demerits/disadvantages of functional organization are given in Table 12.3.

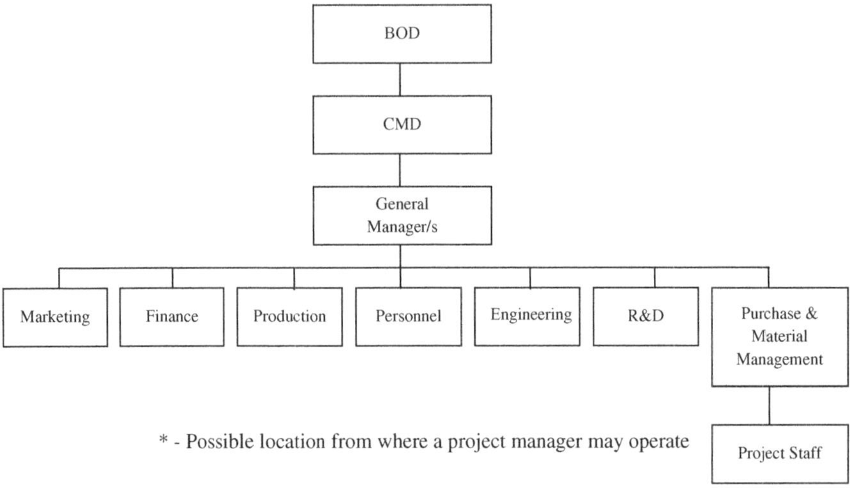


**Fig. 12.4** Functional organization structure

**Table 12.3** Comparison of functional and projectized organizations

| Merits/advantages | Demerits/disadvantages |
|---|---|
| Technical personnel can be used efficiently | Customer interface is weak |
| Good career opportunities for technical personnel likely | Project authority is weak |
| Good stability and security | Horizontal communication is poor |
| Good technology transfer likely between projects | Emphasis more on discipline than programs/ projects |

## 12.9 Matrix Structure

In this structure, the functional and product structures are combined simultaneously at the same level of the organization. For the matrix structure to come into being, in an organization or SBU, the following three conditions must exist:

- Ideas need to be cross-fertilized across projects or products
- Resources are scarce
- Abilities to process information and to make decisions need to be improved [2].

According to the authorities on matrix structure, Davis and Lawrence [3], three distinct phases can be seen in the development of this structure:

1. Temporary cross-functional task forces. This approach is used when a new product is introduced. The project manager is the key link. Boeing makes use of this approach.

2. Products/Brand management. When the cross-functional task force becomes more permanent, the project manager becomes product/service brand manager and the second phase begins. The product/service brand managers act as integrators of semipermanent products or brands. Procter & Gamble's success is attributed to this.
3. Mature matrix. The third and the final phases are true dual-authority structure. Here, both functional and product structure are permanent. The reporting is to a vertical functional superior and a horizontal product manager. Both have equal authority and must work in unison to resolve differences. However, it must be stated that this form has not been very successful.

## 12.10 Projectized Organization

A fully projectized organization is shown in Fig. 12.5. Here, the project manager is in control of all the elements needed for successful execution of project. This structure is suitable for large projects of an organization like ISRO. We have discussed different organizations in Chap. 9.

## 12.11 Comparison

Larson and Gobeli [4] compare the effectiveness of five management structures as under:

1. Functional Organization: The project is divided into segments and assigned to relevant functional areas and/or groups within functional areas. The project is coordinated by the functional and upper levels of management.
2. Functional matrix (similar to functional organization): A person is formally designated to oversee the project across different functional areas. This person has limited authority over functional people involved and serves primarily to plan and coordinate the project. The functional managers retain primary responsibility for their specific segments of the project.
3. Balanced Matrix (similar to matrix organization): A person is assigned to oversee the project and interacts on an equal basis with functional managers. This person and the functional managers jointly direct work flow segments and approve technical and operational decisions.
4. Project Matrix (similar to matrix organization): A manager is assigned to oversee the project and is responsible for the completion of the project. The functional manager's involvement is limited to assigning personnel as needed and providing advisory expertise.

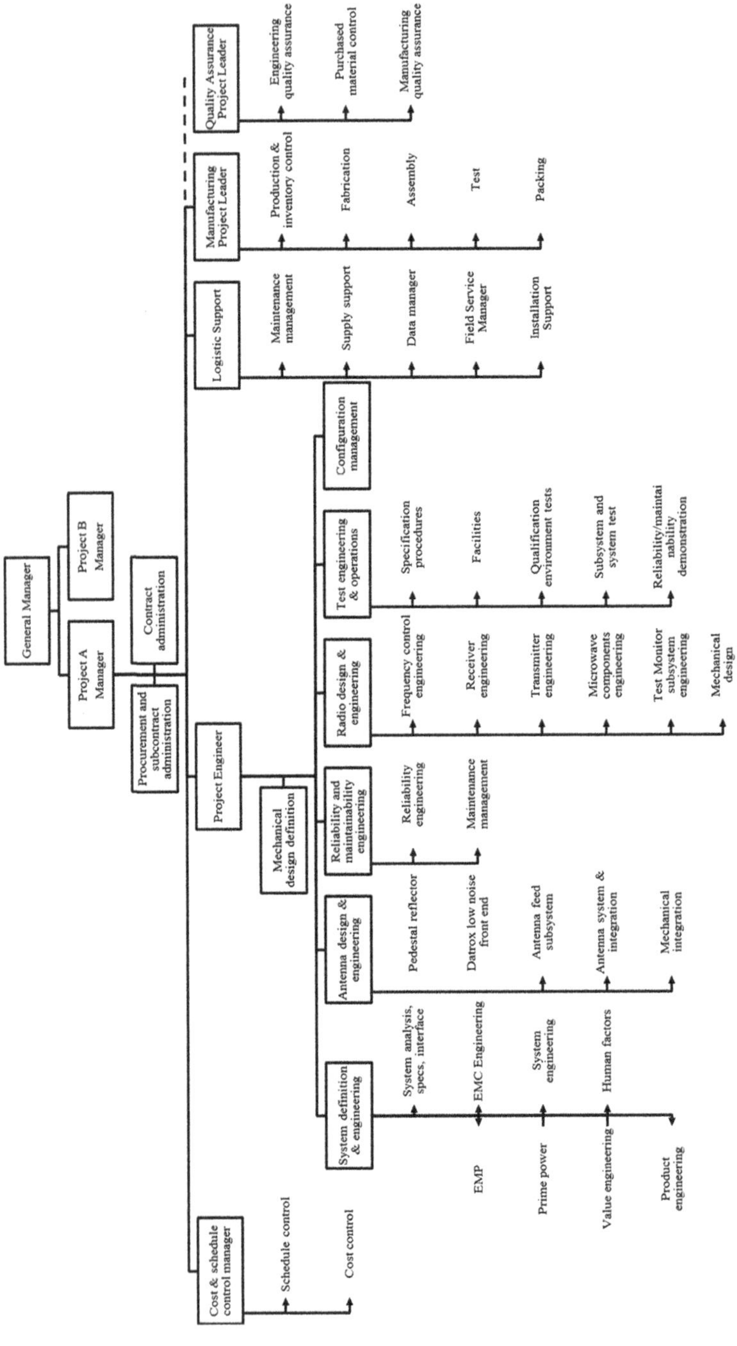


**Fig. 12.5** Fully projectized organization (*Source* [5])

5. Project Team (similar to projectized organization): A manager is put in charge of a project team composed of a core group of personnel from several functional areas and/or groups, assigned on full-time basis. The functional managers have no formal involvement.

Managers engage in risk management activities to compensate for the uncertainty inherent in project management and that things never go according to plan. Risk management is proactive not reactive. It reduces the number of surprises and leads to a better understanding of the most likely outcomes of negative events.

## 12.12 Project Performance

A project manager who is technically and managerially qualified with good interpersonal skills could be a strong motivator for effective performance by all concerned in projects. He can also handle conflict effectively which may be due to: schedules, priorities, resources, procedures, and personality. According to Thamhain and Wilemon [6], the methods most and least favored by one hundred effective project managers are as follows: confrontation (problem solving), compromise (negotiating), smoothing (emphasizing the points of agreement), forcing (one's view point), and withdrawal (retreat in that order).

Good communication and coordination—formal and informal help in monitoring project performance in the customer organization leads to its success. We will detail out the importance of good communication for flawless execution of project in a later chapter.

According to Balderton et al. [7],

> In any development process, the key management task is strategic creation of an early learning path. Project activities are scheduled to gain maximum learning in the early period when cumulative project expenditure is still low. Project controlling soft and time schedules have discussed in Chap. 5. They will help project plans become reality.

## 12.13 Summary

We defined project planning and the characteristics of a project. We clearly differentiated the difference between projects and tasks. We then looked at the influence of the overall strategy of the company on the project. We looked at hoe projects that are planned and the various project planning tools that can be used to measure and monitor the progress of a project. We looked at the various organizational structures and learned what a projectized organization is.

## 12.14 Review Questions

1. Which of the following would be considered a project (a) Construction of a dam across river Cauvery in Karnataka, (b) Setting up of Kaiga nuclear plant, (c) Design of an Indian space shuttle to land on the moon, (d) A production order for supply of spares to a space shuttle?
2. Discuss the project proposal process. How does a successional firm go about preproposal effort for a proposal preparation?
3. Prepare a milestone schedule and WBS for a project of your choice.
4. Develop a project management plan—strategic and implementation plan for a new apparel factory using the concepts discussed in the chapter?
5. In the apparel factory project, discuss the changes needed if the strategic focus of the company shifts from fabrics to fashion clothing.
6. Discuss the merits and demerits of a functional organization.
7. What are the arguments for and against matrix organization?
8. What are the elements of a project-driven organization?
9. Describe the results of the study by Larson and Gobeli on the effectiveness of the five project management structures.
10. What are the characteristics of an effective project manager?
11. What are the methods of conflict management? Which is the most and least effective?
12. What are the keys to a project's success?

## References

1. Kerzner H (1984) Project management's systems approach to planning, scheduling, and controlling, 2nd edn. Van Reinhold Inc, New York, p 823
2. Hrebiniak LG, Joyce WF (1984) Implementing strategy. Macmillan, New York, pp 85–86
3. Davis SM, Lawrence PR (1977) Matrix. Addison-Wesley, Boston, pp 11–24
4. Larson EW, Gobelli DH (1989) Significance of project management structure on development success. In: IEEE transactions on engineering management, p 124
5. Russell D (1976) Archibald, managing high-technology programs and projects. Wiley, New York, pp 104–105
6. Thamhain H, Wilemon DL (1975) The effective management of conflict in project in project-oriented work environment. Defence Manage J, 29–40
7. Balderston J, Birnbaum P, Goodman R, Stahl M (1984) Modern management techniques in engineering and R&D. Van Nostrand Reinhold Company Inc, New York, pp 135–149

# Chapter 13
# Professional Communication

Learning Objectives:

- Understand the importance of good communication
- Understand how to communicate effectively within the organization
- Understand the importance of communication while handling projects.

## 13.1 Introduction

In this chapter, we look at the expanse of communication and its use in preparation of reports and documents. We look at letter writing—internal communication to higher authorities. We also consider comparative statements for a quotation.

## 13.2 Scope and Effectiveness of Communication

Communication is as old as man. It seeks to inform, educate, train, motivate, integrate, relate, promote, entertain, and facilitate decision making. Figure 13.1 gives the different communication methods.

Oral communication can be both structured and unstructured. Written communication is always well structured. Nonverbal communication is neither oral nor written. It thrives on observation—can be body language, gestures, postures, appearance, expressions, etc. Visual communication is through pictures, graphs, charts, signal, and symbols. For example, see (Fig. 13.2).

Audiovisual communication involves demonstration and presentation through audiovisual aids. They are very effective and carry high retention value—ICICI bank ads on TV (Hum Hai Na). In an organization, communication can be upward—employee to supervisor, supervisor to manager, manager to General Manager, and General Manager to Chief Executive. In contrast, downward communication is from Chief Executive to General Manager, General Manager to manager, manager to supervisor, and supervisor to employee.

R. Srinivasan, *Strategic Business Decisions*,
DOI: 10.1007/978-81-322-1901-9_13, 

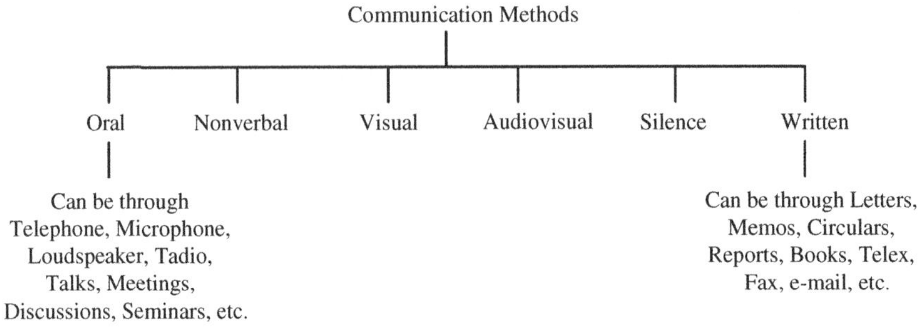


**Fig. 13.1** Communication methods

**Fig. 13.2** Visual communication

There are many definitions of communication—Red fields communication—the broad field of interchange of facts and opinions to the more modern—"Communications maintains and animates life. It creates a common pool of ideas, strengthens the feeling of togetherness through exchange of messages, and translates thoughts into action," UNESCO—many voices in one world.

For communication to be effective, the following are essential: (1) clarity of purpose, (2) process understanding, (3) address the right target, (4) planning, (5) positive approach and sincerity, (6) adequate information, (7) good communication skills, (8) consistency and timeliness, (9) use of appropriate cost-effective modes and channels, and (10) concern for feedback.

## 13.3 Internal Communication

Banks, large companies, and big businesses often have a network of branches, offices, and departments. There can also be several tiers (one to four) of functioning spread from different parts of the country/world. Internal communication covers the following: (1) Inter-department, (2) Intra department, (3) Branch to controlling office, (4) Branch to branch, (5) From sales and field staff to reporting authority, and (6) From functional department to administration.

Circulars and memos represent a common method of conveying specific messages. Circulars may be specific subject-related instructions and memos may cover events and developments. Office notes are sent by sections, departments, or units to higher authorities. It may be put up for orders or for information. Office notes can be put to CEO, CMD, etc., as the case may be.

Letters to higher authorities should carry all relevant information and made in a polite manner. Motivational communication comes from the top management to other levels of employees in the organization. It should be transparent and sincere. A good writer should, in addition to saying "yes" also cultivate the fine art of saying "no." Sending repeated reminders and follow-up letters is not desirable and should be discouraged. Organizations should endeavor to have responsive work culture. Employee newsletters can be an effective medium of communication of information, and developments of relevance can be shared with employees on a periodic basis. All internal communication in an organization should be done selectively and also be well desired.

## 13.4 What Is Communication?

According to Charles Wright, communication is the process of transmitting meaning between individuals.

A more comprehensive definition is given by Edgar Crane who says

> communication is the study of who says what to whom in what settings, by which channel and with what purposes and what effects. It deals with messages designed to influence human behaviour, the media that carry such message, and the markets that respond to such messages.

The question that then needs to be answered is what effective communication is.

Any communication process is efficient to the degree rational judgements are facilitated. We are concerned with communication as a tool or means of developing improved strategies in order that we may better perform the task/project at hand. An efficient communication process should aim to bring about the desired consequence and reduce the undesired effects.

### *13.4.1 The Ideal Communication System*

In a more technical sense, every communication process needs to have an origin and an end and an intervening time and space. In an ideal communication system, the information that began at the origin should be received at the end with absolutely no distortion. For this, the receiver and the sender should be in tune. If not, no communication is possible. The intervening time and space should add little or no noise and should not distort the message. Effective communication

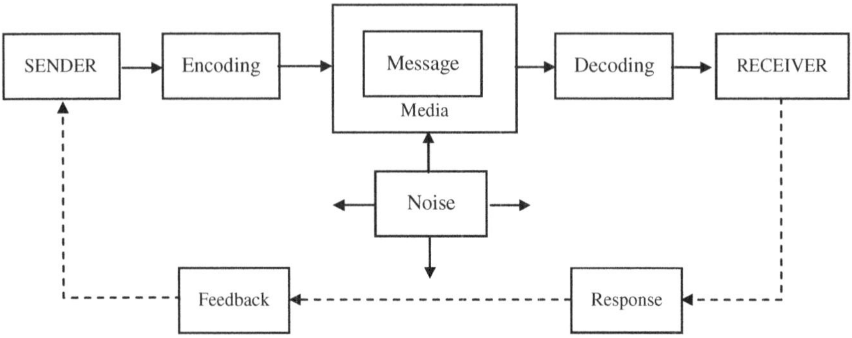


**Fig. 13.3** Ideal communication system

systems also need an effective feedback mechanism that conveys to the sender the receipt of the intended message. It also involves a learning process where both the sender and the receiver incrementally learn how to communicate and understand the intended communication (Fig. 13.3).

Communication also has a communication effect—according to Schramm, these are the result of a number of forces. These forces include the situation, the personality state of the receiver, and the group relations (predispositions of the group that a receiver belongs to). Communicational effect may include the following: Minor attitudinal change and reenforcement, creation of opinions on new issues, and conversion—significant change in thought and behavior.

Factors that influence the communication effect:

- Selective perception
- Selective exposure
- Selective retention.

### 13.4.2 Types of Communication

See (Fig. 13.4).

### 13.4.3 Types of Organizational Communication

See (Fig. 13.5).

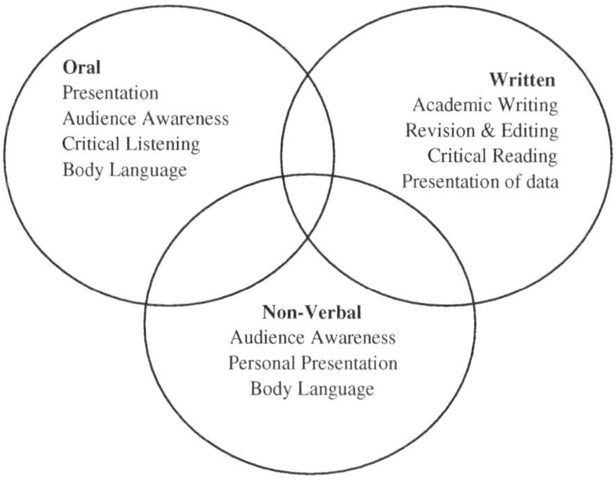


**Fig. 13.4** Types of communication

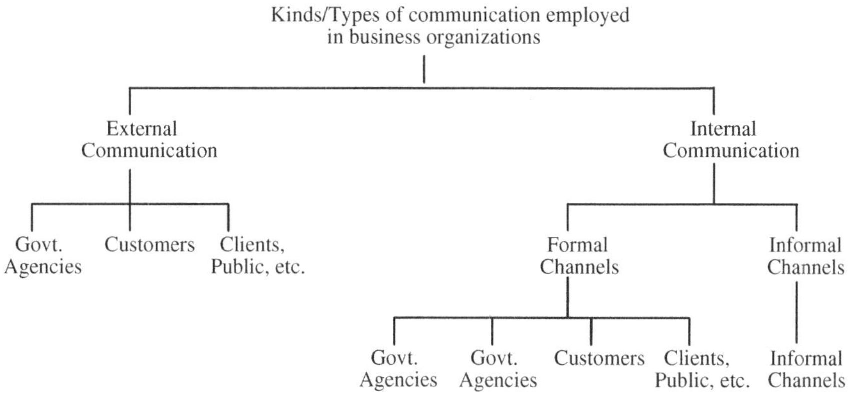


**Fig. 13.5** Organizational communication

## 13.5 Project Communication

Academic reports are used for comprehensive learning—it can be project report, theses, case study, term papers, working papers, etc. Managerial reports are used for decision making—they are brief but comprehensive and reflect the initiator's thinking.

Reports can be as follows: (1) information reports—help understand the existing situation—for business, technology, labor, etc., (2) decision reports—help identify the problem, construct criteria, evaluate alternative options, make a decision, draw action plan, and also work out a contingency plan; (3) research reports—have literature survey to identify knowledge gap, nature and scope of study, methodology, data analysis, findings, conclusions, recommendations, and suggestion for further research.

The question to be addressed now is how to create an effective internal communication? Effective internal communication is one of the key enablers of employee engagement and thought to add significant value to organizations on all metrics from productivity to customer research.

As noted by Quirke: "Traditionally, internal communications has focused on the announcement of management conclusions and the packaging of management thinking into messages for mass distribution to the 'troops'." Research indicates a limit to the value of this "broadcasting" model of IC.

Without feedback loops and harnessing the active involvement and mediation skills of frontline supervisors or team leaders, broadcasting tends to be more effective at influencing senior and middle managers than frontline employees.

The company/the manager communicating internally should have a clear communication plan that covers the following:

1. Situation analysis/background
2. Quantifiable objectives/goals
3. Target audience(s)
4. Key messages
5. Strategies and tactics to meet objectives/goals
6. Responsibility/accountability for completing tactics
7. Budget (as appropriate)
8. Measurement.

An effective internal communication occurs when the sender and the receiver are tuned-in, as in any form of communication with little "channel" disturbance. However, the onus in an organization is to have a very effective top-down communication and a formal channel of feedback systems. The feedback systems need to be open and non-prejudiced. A prejudiced feedback system leads to gossip and grapevine that will stunt any communication.

The manager/CEO also needs to understand the audience and the mood of the audience, and accordingly tailor the message.

In the next section, we discuss the importance of communication in a project team:

Communicating during the first project meeting:

The most important questions to be addressed during the first meeting are as follows:

- The scope and objectives
- The general schedule
- Procedures to be followed
- Introduction of the team members to each other
- Defining the role of each member
- Setting a protocol so that the team members can work with each other with no distractions on the roles and procedures
- Establish a "camaraderie" to ensure the team members work together on the project.

The first meeting and early communication between the project manager and the team set the mood for the rest of the project, and it is important that the communication done during the first meeting is clear and unambiguous. During the course of establishing operational procedures, the project manager should begin working with members to establish the norms for team interaction.

### 13.5.1 Communicating Through the Course of the Project

Though the first meeting defines the requirements and lays the foundation for the future, it is important for the project manager to communicate effectively to the team throughout the execution and management of the project. The project manager should also ensure that the team communicates well with each other. This should be done in formal meetings and informally while executing the project. Examples of subsequent meetings could include status report meetings, problem-solving meetings, and audit meetings. Meetings are often considered an anathema to productivity. The most common complaint is that meetings last too long. Establishing an agenda and adjournment time helps participants, budget discussion time and provides a basis for expediting the proceedings. Apart from reviewing the progress and trying to solve problems, project meetings, certainly in the initial stages, should aim to create a shared vision and this vision should align with the overall strategic plan of the organization.

### 13.5.2 Communication to Manage Conflicts

During project definition, the most significant sources of conflict are priorities, administrative procedures, schedule, and workforce. Disputes occur over the relative importance of the project compared with other activities, which project management structure to use (especially how much control the project manager should have), the personnel to be assigned, and the scheduling of the project into existing workloads. During the execution phase, friction arises over schedule slippage, technical problems, and staff issues. Milestones become more difficult to meet because of accumulating schedule slippages. During the delivery phase, schedules continue as the biggest source of conflict as schedule slippages make it more difficult to meet target completion dates (Source: Erik Larson and Clifford Grey. Project Management, 5th edition).

The shared vision established during the initial meeting will help the project teams overcome these issues. It also becomes imperative for the project manager to "step outside" of the conflicting situation and understand the source of the problem and take decisions based on the overall goals to be achieved by the project.

While it is important to communicate to resolve conflicts, it is also important that the team is rewarded and appreciated for a job done well.

### *13.5.3 Preparation of Reports*

Reports are a must for tracking the progress of the project. Reporting, though cumbersome, is a must to ensure the successful delivery of the project. A weekly, monthly, and quarterly report depending on the project timelines help both the team and the organizations' top management understand the progress and the lacunae of the projects.

The final report, though specific to a project and the organization, will largely consist of the following:

- Executive summary
- Review and analysis
- Recommendations
- Lessons learned
- Appendix.

## 13.6 Comparative Statements of a Profession

An inquiry has to be first generated specifying what is wanted—for example, if it is a laptop—what are the specifications under various levels of performance, etc. The inquiry gives the last date for quotation. Quotations are generally received from three suppliers (for enquiries above Rs. 25,000/-) before a particular date. These are compiled and compared against the various heads of inquiry. A comparative statement is then made. Recommendation is normally made for the lowest quote unless there are compelling reasons for not doing so.

**Inquiry Note**

Please submit your quotation for the following item, in the same sequence and format by 1600 hrs on December 12, 2005.

Supply of laptop: 1 No. by December 18, 2005
Specifications for laptop

| Description | Intended specs (Laptop) | Suppliers specs Make/ Model# | Quoted price |
|---|---|---|---|
| Processor | Intel technology—Intel Celeron/Pentium-M processor @ 1.5 GHz or above | | |
| Memory | 256 MB DDR SDRAM expandable to 2048 MB | | |
| Display | 14" XGA TFT display | | |

(continued)

(continued)

Supply of laptop: 1 No. by December 18, 2005
Specifications for laptop

| Description | Intended specs (Laptop) | Suppliers specs Make/ Model# | Quoted price |
|---|---|---|---|
| Hard disk | 40 GB or above | | |
| Multimedia | DVD-CDRW Combo Drive (max 24x when CD-ROM is accessed) | | |
| Graphics | 4x AGP bus, up to 64 MB shared memory | | |
| Keyboard | Ergonomic keyboard | | |
| Communication | Integrated V.90/56K software modem, 10/100 Mbps Ethernet | | |
| Wireless communication | Integrated IR, Integrated wireless LAN | | |
| Operating system | Windows XP preloaded with recovery disk | | |
| Battery | Rechargeable Lithium ion battery, 3.5 hours or more of battery life | | |
| Weight | Less than 3 Kg (with battery and optical disk drive) | | |
| Warranty | 3-year warranty | | |
| Audio | Built-in 16-bit stereo speakers with volume control | | |
| Expansion slots | Parallel, RGB, USB 2.0 x 2, RJ-11, RJ-45, i-Link, S-video TV-out, DC-in, etc. | | |
| Floppy drive | Built-in/USB Floppy Drive | | |
| Optical mouse | Optical mouse and mouse pad | | |
| Freight and installation | Free delivery and installation at IISc, Bangalore | | |
| Optional | Additional cost for 512 MB in lieu of 256 MB | | |
| Additional accessories | Leather carry case with external scroll mouse | | |
| Other details, if any | | | |

## Recommendations

**Sub**: Recommendations for the Procurement of Laptop

There is a budget provision of Rs. 35,000 towards procurement of equipment and accordingly, enquiries as per attachment are sent to the suppliers for purchase of Laptop, required for carrying out the project.

Enquiries are sent to five suppliers and Quotations were received from four suppliers. The comparative statement of the four offers received is enclosed.

Out of these, the technically suitable L1 offers are from Supplier 2 with Rs. 35,000 as the basic system cost + 4 % VAT of Rs. 1,400. However, as the budget provision is only for Rs. 35,000 towards this purchase, the price was negotiated with the supplier to reduce the price to an all-inclusive price of Rs. 35,000 which is reflected in the quotation.

Table 13.1 Comparative Statement of competing quotes for Laptop

| Item/Feature | Intended specs (Laptop) | 1 Supplier specs | Quoted price | 2 Supplier specs | Quoted price | 3 Supplier specs | Quoted price | 4 Supplier specs | Quoted price |
|---|---|---|---|---|---|---|---|---|---|
| Make/Model | | | | | | | | | |
| Processor | Intel technology—Intel Celeron/ Pentium-M processor @ 1.5 GHz or above | Yes | | Presario | 35000 | ACER TM 2413LMi | 42120 | Compac Presario V 23474 AP | 39490 |
| Memory | 256 MB DDR SDRAM expandable to 2048 MB | Yes | | Celeron M2202 | | Yes | | Celeron M360 | |
| Display | 14" XGA TFT display | Yes | | Yes | | Yes | | Yes | |
| Hard disk | 40 GB or above | Yes | | Yes | | Yes | | Yes | |
| Multimedia | DVD-CDRW Combo Drive (max 24x when CD-ROM is accessed) | Yes | | Yes | | DVD writer | | Yes | |
| Graphics | 4x AGP bus, up to 64 MB shared memory | Yes | | Yes | | Yes | | Yes | |
| Keyboard | Ergonomic keyboard | Yes | | Yes | | Yes | | Yes | |
| Communication | Integrated V.90/56K software modem, 10/100 Mbps Ethernet | Yes | | Yes | | Yes | | Yes | |
| Wireless communication | Integrated IR, Integrated wireless LAN | Yes | | No | | Yes | | No | |
| Operating system | Windows XP preloaded with recovery disk | Yes | | Yes | | Yes | | Yes | |
| Battery | Rechargeable Lithium ion battery, 3.5 hours or more of battery life | Yes | | Yes | | Yes | | Yes | |
| Weight | Less than 3 Kg (with battery and optical disk drive) | Yes | | Yes | | Yes | | Yes | |
| Warranty | 3-year warranty | Yes | | 1 year | | 1 year | | 1 year | |

(continued)

**Table 13.1** (continued)

| Item/Feature | Intended specs (Laptop) | 1 Supplier specs | Quoted price | 2 Supplier specs | Quoted price | 3 Supplier specs | Quoted price | 4 Supplier specs | Quoted price |
|---|---|---|---|---|---|---|---|---|---|
| Make/Model | | | | | | | | | |
| Audio | Built in 16-bit stereo speakers with volume control | Yes | | Yes | | Yes | | Yes | |
| Expansion slots | Parallel, RGB, USB 2.0 x 2, RJ-11, RJ-45, i-Link, S-video TV-out, DC-in, etc | Yes | | Yes | | Yes | | Yes | |
| Floppy drive | Built-in/USB Floppy Drive | Yes | | Yes | | USB | | USB | |
| Optical mouse | Optical mouse and mouse pad | Yes | | Yes | | | | No | |
| Freight and installation | Free delivery and installation at IISc, Bangalore | Yes | | Yes | | | | Yes | |
| Optional | Additional cost for 512 MB in lieu of 256 MB | Yes | | Yes | | | | Yes | |
| External mouse and pad | | Yes | | Yes | | | | Yes | |
| Additional accessories | | | | | | | | | |
| | TOTAL COST | | 41000 | | 44400 | | 42120 | | 41070 |
| | System cost | | 4100 | | 35000 | | 42120 | | 41070 |
| | Status | | | | L1 | | | | |
| | VAT | | Extra 4% | | Extra 4% | | Extra 4% | | Extra 4% |
| | System cost + VAT | | 42460 | | 36400 | | 43805 | | 42712 |

Hence it is recommended to place the order on Supplier 2 as per the specifications enclosed. The payment terms may include "100 % against delivery and acceptance."

## 13.7 Summary

In this chapter, we reviewed the basics of communication—the sender and the receiver, the context of the message, the background noise that could distort the message, and the channel through which communication can happen. We understood effective communication has "communication effects" and we looked at factors that have an impact on the communication effect. We looked at the types of organizational communication.

In the subsequent section, we discussed the importance of effective communication and the importance of communication for project management. We then looked at the communication requirement at different stages of project management.

## 13.8 Discussion Questions

1. What do you understand by communication? What are the various types of communication?

2. What is internal communication? How can internal communication be effective?

3. What is the importance of reports for project management?

4. The project team working on the company's knowledge management program needs to be informed that the project will no longer receive funds due to the company's cost-cutting efforts. The team has completed 60 % of the project and is ahead of schedule. Draw out a communication plan to inform the team of shelving of the project and the retrenchment of a few people and the redeployment of the rest to other teams.

5. Create a communication plan for a project team that is responsible for setting up a new production line of chips in an existing factory.

# References for Further Reading

Bazaraa M et al (1990) Linear programming and network flows, 2nd edn. Wiley, New York
Chelst KR, Edwards TG (2004) Does this line ever move? Everyday applications of operations research. Key Curriculum Press, Berlin
Chvatal V (1983) Linear programming. W.H. Freeman Publications, New York
Dantzig GB, Thapa MN (1997) Linear programming I: Introduction. Springer, New York
DiSanza JR, Legge NJ (2012) Business and professional communication: plans, processes, and performance. Pearson, London
Dixit AK (1999) Games of strategy. W. W. Norton & Company, New York
Dixit AK, Nalebuff BJ (1991) Thinking strategically: the competitive edge in business, politics, and everyday life. W. W. Norton & Company, New York
Eiselt HA, Sandblom C-L (2010) Operations Research: a model-based approach. Springer, New York
Evans JR, Minieka E (1992) Optimization algorithms for networks and graphs, 2nd edn. Marcel Dekker, New York
Ficken FA (1961) The simplex method of linear programming. Holt, Rinehart and Winston, New York
Fourer R, Gay DM, Kernighan BW (2002) AMPL: a modeling language for mathematical programming (chapter 3). Duxbury Press/Brooks/Cole Publishing Company, North Scituate
Gass SI (2003) Linear programming: methods and applications. Courier Dover Publications, New York
Gen M, Cheng R, Lin L (2008) Network models and optimization: multiobjective genetic algorithm approach. Springer, New York
Gintis H (2009) Game theory evolving: a problem-centered introduction to modeling strategic interaction. Princeton University Press, Princeton
Gross D, Shortle JF, Thompson JM, Harris CM (2008) Fundamentals of queueing theory, 4th edn. Wiley, New York
Haghighi AM, Mishev DP (2008) Queuing models in industry and business. Nova Publishers, New York
Koneru A (2008) Professional communication. Tata McGraw-Hill Education, New Delhi
Kuhn HW (2010) The Hungarian method for the assignment problem, 50 years of integer programming 1958–2008. Springer, New York
Maros I (2003) Computational techniques of the simplex method. Springer, New York
Murty K (1992) Network programming. Prentice Hall, Upper Saddle River
Sinha SM (2006) Mathematical programming (chapter 20). Elsevier Science, Amsterdam
Tulsian PC, Pandey V (2006) Quantitative techniques: theory and problems. Pearson Education India, New Delhi
Vanderbei RJ (1997) Linear programming: foundations and extensions. Springer Publications, New York

R. Srinivasan, *Strategic Business Decisions*,
DOI: 10.1007/978-81-322-1901-9, © Springer India 2014

von Neumann J, Morgenstern O (2007) Theory of games and economic behavior. Princeton University Press, Princeton

Wagner HM (1975) Principles of operations research: with applications to managerial decisions. Prentice Hall, Englewood Cliffs

Whalen DJ, Ricca TM (2006) The professional communications toolkit. Sage Publications, New York

# Index



R. Srinivasan, *Strategic Business Decisions*,
DOI: 10.1007/978-81-322-1901-9, © Springer India 2014

The manufacturer's authorised representative in the EU is Springer Nature Customer Service Centre GmbH, Europaplatz 3, 69115 Heidelberg, Germany. If you have any concerns regarding our products, please contact ProductSafety@springernature.com

Printed and bound by CPI Group (UK) Ltd, Croydon, CR0 4YY
30/09/2026
02225447-0009